Contents

1 Fundamental Concepts and Notation   **13**
    Variables 13
    Constants 13
    Algebraic Expressions 14
    Equations 14
    Multiple Choice Questions 14
    Practice Problems 17
    Answers 19

2 Real Numbers and Their Properties   **21**
    The Real Number System 21
    1   Classification of Real Numbers 21
    2   Representation on the Number Line 21
    Properties of Arithmetic Operations 22
    1   The Commutative Property 22
    2   The Associative Property 22
    3   The Distributive Property 23
    4   The Closure Property 23
    5   Identity and Inverse Properties 23
    Multiple Choice Questions 24
    Practice Problems 26
    Answers 28

3 Order of Operations and Expression Evaluation   **32**
    The Hierarchy of Operations 32
    1   Parentheses and Grouping Symbols 32
    2   Exponents and Root Operations 32
    3   Multiplication and Division 33
    4   Addition and Subtraction 33
    Evaluation of Algebraic Expressions 33

	Strategies for Methodical Evaluation	34
	Multiple Choice Questions	34
	Practice Problems	37
	Answers	39

4 Combining Like Terms and the Distributive Property 43

Identifying Like Terms	43
Combining Like Terms in Algebraic Expressions . . .	43
The Distributive Property and Its Applications . . .	44
1 Intertwining Distribution with Combination of Like Terms	45
Multiple Choice Questions	45
Practice Problems	48
Answers	50

5 Exponents and Powers: Concepts and Applications 54

Definition and Notation	54
Multiplication of Exponential Terms	54
Division of Exponential Terms	55
The Power of a Power Rule	55
Application to Algebraic Expressions	56
Multiple Choice Questions	56
Practice Problems	58
Answers	60

6 Scientific Notation and Special Products 64

Scientific Notation	64
Special Products	65
1 Square of a Binomial	65
2 Product of a Sum and a Difference	66
Multiple Choice Questions	66
Practice Problems	68
Answers	71

7 Factoring: Greatest Common Factor and Simple Patterns 73

Extracting the Greatest Common Factor	73
Recognizing Basic Factoring Patterns	74
Worked Examples	75
Multiple Choice Questions	76
Practice Problems	78

Answers 80

8 Factoring: Difference of Squares and Perfect Square Trinomials 84
Difference of Squares 84
Perfect Square Trinomials 85
Multiple Choice Questions 86
Practice Problems 88
Answers 90

9 Factoring: Grouping and Quadratic Trinomials 92
Factoring by Grouping 92
Factoring Quadratic Trinomials 93
Multiple Choice Questions 95
Practice Problems 97
Answers 99

10 Solving Linear Equations: Basic Techniques 102
Definition and Structure of One-Variable Linear Equations 102
Properties of Equality and Inverse Operations 102
Techniques for Isolating the Variable 103
Step-by-Step Examples 104
1   Example 1 104
2   Example 2 104
3   Example 3 104
Verification of Derived Solutions 105
Multiple Choice Questions 105
Practice Problems 107
Answers 109

11 Solving Linear Equations with Variables on Both Sides 113
Structure and Characteristics of Two-Sided Equations 113
Fundamental Properties and Principles of Equality . 114
Methodical Approach for Isolating the Variable . . . 114
Worked Examples 115
1   Example 1 115
2   Example 2 115
3   Example 3 116
Verification of Solutions 116
Multiple Choice Questions 116

        Practice Problems 119
        Answers 121

12 Applications of Linear Equations — 125
        Translating Real-Life Scenarios into Mathematical
        Models 125
        1   Identification of Variables and Constants . . 125
        2   Constructing the Algebraic Equation 126
        Algebraic Techniques in Solving Word Problems . . 126
        1   Application of Inverse Operations 126
        2   Verification and Interpretation of Solutions . 127
        Illustrative Examples of Practical Applications . . . 127
        1   Applications in Financial Calculations 127
        2   Applications in Rate Problems 128
        3   Applications in Mixture Problems 128
        4   Integration of Analytical Methods 128
        Multiple Choice Questions 128
        Practice Problems 131
        Answers 133

13 Solving Linear Inequalities — 136
        Definition and Fundamental Concepts 136
        1   Inequality Symbols and Notation 136
        2   Relationship Between Linear Equations and
             Inequalities 136
        Techniques for Solving Linear Inequalities 137
        1   Isolating the Variable 137
        2   Application of Inverse Operations 137
        3   Handling Negative Coefficients 137
        Graphical Representation on the Number Line . . . 138
        1   Concept of the Number Line 138
        2   Endpoints: Open and Closed Circles 138
        3   Shading the Interval 138
        Worked Examples with Step-by-Step Explanations . 138
        Multiple Choice Questions 139
        Practice Problems 142
        Answers 144

14 Compound Inequalities and Absolute Value Equations — 147
        Compound Inequalities 147

1	Definition and Notation of Compound Inequalities	147
2	Techniques for Solving Compound Inequalities	148
3	Worked Examples on Compound Inequalities	148

Absolute Value Equations 149
1	Definition and Properties of Absolute Value	. 149
2	Solving Equations Involving Absolute Values	149
3	Worked Examples on Absolute Value Equations	150

Multiple Choice Questions 150
Practice Problems 153
Answers 155

15 Graphing on the Cartesian Plane   159
The Cartesian Coordinate System 159
Plotting Points on the Cartesian Plane 160
Geometric Interpretation of Algebraic Data 160
Multiple Choice Questions 161
Practice Problems 163
Answers 165

16 Understanding Slope and Intercepts   168
The Concept of Slope 168
The y-Intercept: Identifying the Vertical Intersection 169
The x-Intercept: Identifying the Horizontal Intersection 169
Analyzing Linear Relationships Through Slope and Intercepts 170
Multiple Choice Questions 170
Practice Problems 172
Answers 174

17 Graphing Linear Functions   177
Graphing with Slope-Intercept Form 177
Graphing with Point-Slope Form 178
Interpreting the Behavior of Linear Functions 178
Multiple Choice Questions 179
Practice Problems 181
Answers 183

18 Graphing Linear Inequalities — 187
- Understanding Linear Inequalities 187
- Plotting the Boundary Line 187
- Determining the Solution Region 188
- Worked Example in Graphing Inequalities 188
- Multiple Choice Questions 189
- Practice Problems 192
- Answers 195

19 Solving Systems of Linear Equations — 198
- Overview of Systems of Equations 198
- Method of Substitution 198
  - 1 Worked Example: Substitution Method ... 199
- Method of Elimination 199
  - 1 Worked Example: Elimination Method 200
- Systems with More Than Two Variables 201
- Multiple Choice Questions 201
- Practice Problems 204
- Answers 206

20 Applications of Systems of Equations — 212
- Model Formulation and Problem Analysis 212
  - 1 Interpreting Variables and Constraints 212
  - 2 Constructing Systems from Real-World Data 213
- Worked Example: Production Scheduling in Manufacturing 213
  - 1 Problem Description and System Setup ... 213
  - 2 Solution Process Using the Elimination Technique 214
- Worked Example: Mixture Problem in Chemical Solutions 214
  - 1 Problem Description and Formulation 215
  - 2 Solution Process Using the Elimination Method 215
- Multiple Choice Questions 216
- Practice Problems 218
- Answers 222

21 Polynomials: Terminology and Degree — 225
- Definition and Fundamental Concepts of Polynomials 225
- Terms and Structure in Polynomials 225
- Degree of a Polynomial 226
- Coefficients and Their Roles 226

Multiple Choice Questions	227
Practice Problems	229
Answers	231

22 Operations with Polynomials   234

Addition of Polynomials	234
Subtraction of Polynomials	235
Multiplication of Polynomials	235
Multiple Choice Questions	236
Practice Problems	239
Answers	241

23 Special Products of Polynomials   245

Square of a Binomial	245
Product of a Sum and a Difference	246
Cube of a Binomial	246
Sum and Difference of Cubes	247
Multiple Choice Questions	247
Practice Problems	250
Answers	252

24 Factoring Techniques for Higher-Degree Polynomials   255

Factoring by Grouping	255
Pattern Recognition in Polynomial Factoring	256
Factoring by Substitution	258
Multiple Choice Questions	259
Practice Problems	262
Answers	264

25 Division of Polynomials   268

Polynomial Long Division	268
1 Detailed Process and Example	268
Synthetic Division	270
1 Worked Example	270
Multiple Choice Questions	271
Practice Problems	274
Answers	276

26 Rational Expressions: Concepts and Simplification 284
    Definition and Fundamental Properties 284
    Domain Restrictions and Excluded Values 285
    Simplification Techniques Through Factoring and Cancellation 285
    Multiple Choice Questions 287
    Practice Problems 290
    Answers 292

27 Operations with Rational Expressions 295
    Addition and Subtraction of Rational Expressions . 295
    Multiplication of Rational Expressions 296
    Division of Rational Expressions 296
    Multiple Choice Questions 297
    Practice Problems 300
    Answers 302

28 Solving Rational Equations 306
    Identifying Domain Restrictions 306
    Clearing the Denominators Using the Least Common Denominator 306
    Solving the Resultant Equation 307
    Verification of Solutions and Identification of Extraneous Roots 308
    Multiple Choice Questions 308
    Practice Problems 311
    Answers 313

29 Radical Expressions and Their Properties 318
    Definition and Notation 318
    Fundamental Properties of Radicals 318
    Techniques for Simplifying Radical Expressions . . . 319
    Operations Involving Radical Expressions 320
    Rationalizing the Denominator 320
    Multiple Choice Questions 321
    Practice Problems 323
    Answers 325

30 Operations and Equations Involving Radicals 328
    Basic Concepts and Simplification of Radical Expressions 328
    Arithmetic Operations on Radical Expressions . . . 329

1	Addition and Subtraction of Radical Expressions	329
2	Multiplication of Radical Expressions	329

Solving Equations Involving Radical Expressions . . 330
Multiple Choice Questions 331
Practice Problems 333
Answers 335

31 Quadratic Equations: Forms and Strategies 339

Forms of Quadratic Equations 339
Strategies for Solving Quadratic Equations 340

1	Factoring Methods	340
2	Completing the Square	340
3	Quadratic Formula	341
4	Analysis of the Discriminant	341

Applications of Algebraic Methods 341
Multiple Choice Questions 342
Practice Problems 344
Answers 346

32 Solving Quadratic Equations by Factoring 351

Quadratic Equations in Standard Form 351
Methods for Factoring Quadratic Expressions 351

1	Extracting Common Factors	352
2	Splitting the Middle Term	352
3	Recognition of Special Patterns	353

Application of the Zero Product Property 353
Worked Examples 353

1	Example 1	353
2	Example 2	354
3	Example 3	355

Multiple Choice Questions 355
Practice Problems 358
Answers 360

33 Completing the Square and the Quadratic Formula 364

Completing the Square 364

1	Theory and Method	364
2	Worked Example	365

The Quadratic Formula 366

1	Derivation from Completing the Square . . .	366
2	Interpretation and Applications	367

    Multiple Choice Questions 367
    Practice Problems 370
    Answers 372

34 Mathematical Modeling Using Algebra   377
    Foundations of Algebraic Modeling 377
    Defining Variables and Formulating Relationships . . 377
    Developing Equations from Real-World Contexts . . 378
    Methods for Solving Algebraic Models 378
    Interpreting the Results of Algebraic Models 379
    Detailed Examples in Algebraic Modeling 379
    Multiple Choice Questions 380
    Practice Problems 382
    Answers 385

35 Concept of Functions and Function Notation   388
    Definition and Fundamental Characteristics of Functions 388
    Domain of a Function 388
    Range of a Function 389
    Proper Function Notation 389
    Multiple Choice Questions 390
    Practice Problems 392
    Answers 394

36 Graphing and Analyzing Functions   397
    Fundamentals of Graphing Functions 397
    Transformations of Functions 397
    Intercepts and Key Graphical Features 398
    Analyzing the Overall Shape of Functions 399
    Multiple Choice Questions 400
    Practice Problems 402
    Answers 404

37 Linear Functions and Their Applications   408
    Definition and Fundamental Characteristics 408
    Graphical Representation and Analysis 408
    Algebraic Forms of Linear Equations 409
    Practical Applications in Problem Solving 409
    Interpreting Slope and Y-Intercept in Real-World Contexts 410
    Multiple Choice Questions 410

    Practice Problems 413
    Answers 415

38 Operations and Composition of Functions   417
    Arithmetic Operations on Functions 417
    1   Addition and Subtraction of Functions 417
    2   Multiplication and Division of Functions . . . 418
    Composition of Functions 418
    Multiple Choice Questions 419
    Practice Problems 422
    Answers 425

39 Exponential Functions: Properties and Graphs   429
    Definition and Basic Structure 429
    Growth and Decay Properties 429
    Graphing Techniques for Exponential Functions . . . 430
    1   Identification of Key Characteristics 430
    2   Constructing the Graph 430
    3   Asymptotic Behavior and Domain Considerations 431
    Multiple Choice Questions 431
    Practice Problems 433
    Answers 436

40 Logarithmic Functions and Their Properties   440
    Definition and Inverse Relationship with Exponential Functions 440
    Fundamental Laws Governing Logarithms 441
    1   The Product Rule 441
    2   The Quotient Rule 441
    3   The Power Rule 441
    4   The Change-of-Base Formula 442
    Graphical Characteristics of Logarithmic Functions . 442
    Algebraic Manipulations Involving Logarithms . . . 442
    Multiple Choice Questions 443
    Practice Problems 445
    Answers 447

41 Solving Exponential and Logarithmic Equations   451
    Solving Exponential Equations 451
    1   Exponential Equations with a Common Base 451
    2   Exponential Equations Requiring Logarithms 452

Solving Logarithmic Equations 453
    1   Combining and Isolating Logarithmic Expressions 453
    2   Removing the Logarithm through Exponentiation 453
Equations Involving Both Exponential and Logarithmic Terms 454
Multiple Choice Questions 455
Practice Problems 458
Answers 460

42 Sequences and Algebraic Patterns   464

Arithmetic Sequences 464
    1   Definition and Basic Properties 464
    2   Determining the nth Term 464
    3   Recursive Definitions in Arithmetic Sequences 465
Geometric Sequences 465
    1   Definition and Fundamental Characteristics . 465
    2   Analyzing the nth Term 465
    3   Recursive Formulations in Geometric Sequences 466
Algebraic Patterns and Recursive Definitions 466
    1   Recognition and Analysis of Algebraic Patterns 466
    2   Establishing Recursive Definitions 466
    3   Examples of Recursive Patterns in Algebra . 467
Multiple Choice Questions 467
Practice Problems 469
Answers 471

Chapter 1

Fundamental Concepts and Notation

Variables

Variables serve as placeholders for quantities that are not yet specified or that may vary within a mathematical context. Represented by letters or symbols, variables allow for the formulation of generalized statements and relationships. In algebra, symbols such as x, y, and z are commonly employed to denote these undetermined quantities. For example, in the expression $x+5$, the symbol x represents a number whose specific value is not fixed, enabling the expression to encompass a range of potential numerical situations. The use of variables permits the abstraction and generalization fundamental to algebraic reasoning.

Constants

Constants are fixed numerical values that remain unchanged within a given expression or equation. They provide a stable reference point amidst the variability of algebraic symbols. Examples of constants include numbers such as 2, -3, and even well-known irrational numbers like π. In an expression like $2x+8$, the number 8 performs the role of a constant, anchoring the expression with its invariant value. The distinction between constants and variables is essential, as it allows for the clear identification of elements that are subject to change versus those that maintain consistency throughout the process of algebraic manipulation.

Algebraic Expressions

An algebraic expression is a combination of variables, constants, and arithmetic operations structured according to the rules of algebra. Such expressions may involve addition, subtraction, multiplication, division, and exponents to represent quantities succinctly and powerfully. For instance, the expression $3x - 4$ comprises a term with a variable and an accompanying constant term, while a more elaborate expression like $2x^2 + 5x - 7$ illustrates the assembly of multiple terms with varying degrees. The arrangement of coefficients, variables, and constants in an expression adheres to a systematic order, establishing a coherent structure that facilitates manipulation, simplification, and further analysis.

Equations

Equations articulate the equality between two algebraic expressions through the placement of an equal sign (=) between them. Every equation asserts that, for a particular substitution of values, the expressions on either side of the equal sign yield the same result. Consider the linear equation $2x + 3 = 7$: this statement declares that there exists a value for x such that when twice that value is added to 3, the sum equals 7. The process of solving an equation involves applying operations—while preserving equality—to isolate the variable and determine its precise value. This systematic approach relies on the fundamental properties of equality and the operational laws of arithmetic, ensuring that the transformations performed during the solution process maintain the intrinsic balance prescribed by the equation.

Multiple Choice Questions

1. Which of the following best defines a variable in an algebraic context?

    (a) A fixed number whose value never changes.

(b) A symbol that represents an unspecified or changeable quantity.

(c) An arithmetic operation used for combining numbers.

(d) A sign that indicates the end of an equation.

2. In the expression $2x + 8$, what role does the number 8 play?

    (a) It is a variable.

    (b) It is the coefficient of x.

    (c) It is a constant.

    (d) It is an exponent.

3. Which of the following is an example of an algebraic expression?

    (a) $5 + 3 = 8$

    (b) $2x + 5$

    (c) $x = 4$

    (d) 7

4. In the equation $2x + 3 = 7$, which step correctly isolates the term containing x?

    (a) Add 3 to both sides to obtain $2x = 10$.

    (b) Subtract 3 from both sides to obtain $2x = 4$.

    (c) Divide both sides by 2 immediately to obtain $x + 1.5 = 3.5$.

    (d) Multiply both sides by 2 to obtain $4x + 6 = 14$.

5. What is the fundamental difference between an algebraic expression and an equation?

    (a) An algebraic expression always includes an equal sign, whereas an equation does not.

    (b) An equation contains an equal sign to show equality between two expressions, while an algebraic expression does not necessarily include one.

    (c) Both always contain an equal sign, but only equations involve variables.

    (d) There is no difference; the terms are interchangeable.

6. In the expression 3x - 4, what are the coefficient and constant term, respectively?

    (a) 3 is the coefficient; -4 is the constant.
    (b) x is the coefficient; 3 is the constant.
    (c) 3x is the coefficient; 4 is the constant.
    (d) -4 is the coefficient; 3 is the constant.

7. Why are variables essential in algebraic reasoning?

    (a) They provide fixed numerical values for calculations.
    (b) They allow us to generalize relationships and represent unknown or changeable quantities.
    (c) They eliminate the need to use constants in expressions.
    (d) They ensure that every equation has a unique solution.

Answers:

1. **B: A symbol that represents an unspecified or changeable quantity**
Variables serve as placeholders for values that can vary, enabling the formulation of general statements and relationships in algebra.

2. **C: It is a constant**
In the expression $2x + 8$, the number 8 is fixed and does not change, making it a constant term.

3. **B: $2x + 5$**
An algebraic expression is a combination of variables, constants, and operations. Option B includes both a variable term ($2x$) and a constant (5), illustrating this concept.

4. **B: Subtract 3 from both sides to obtain $2x = 4$**
To isolate the term with x, subtract 3 from both sides of the equation, which correctly yields $2x = 4$ before solving for x.

5. **B: An equation contains an equal sign to show equality between two expressions, while an algebraic expression does not necessarily include one**
This distinction is fundamental: equations use the "=" sign to assert that two expressions are equal, whereas an algebraic expression is simply a combination of symbols and numbers.

6. **A: 3 is the coefficient; -4 is the constant**
   In 3x - 4, the number 3 multiplies the variable x (making it the coefficient), and -4 is the fixed number or constant term.

7. **B: They allow us to generalize relationships and represent unknown or changeable quantities**
   Variables are essential because they enable mathematicians to create generalized models and solve problems where specific numerical values are not initially known.

Practice Problems

1. Identify and classify the terms in the following algebraic expression:
$$5x + 3$$

2. Define the terms "variable" and "constant" in algebra. Explain how each is used when forming algebraic expressions and equations.

3. Construct an algebraic expression to represent the following statement: "The sum of three times a number y and 7."

4. Solve the linear equation by isolating the variable:
$$4x - 5 = 11$$

5. Explain the importance of the equal sign in an equation using the example:
$$2x + 3 = 7$$
In your explanation, discuss what the equal sign signifies with respect to maintaining balance between the two sides.

6. In the algebraic expression:
$$3x^2 - 4x + 8$$
identify the coefficients, the variable parts, and the constant. Provide a detailed explanation for each part of the expression.

Answers

1. **Solution:** The expression
$$5x + 3$$
consists of two terms. In the term **5x**,

   - **5** is the coefficient, which indicates the multiplicative factor of the variable.
   - **x** is the variable, representing an unknown or changeable quantity.

   The term **3** is a constant because it is a fixed numerical value. Thus, in this expression, **x** is the variable; **5** is the coefficient of the variable term; and **3** is the constant term.

2. **Solution:** A **variable** is a symbol (usually a letter such as x, y, or z) that represents an unknown or changeable quantity. Variables allow us to express general relationships and can take on different values. A **constant** is a fixed value that does not change throughout a particular mathematical discussion or problem. In algebraic expressions and equations, variables provide flexibility and generality, while constants anchor the expressions with specific numerical values.

3. **Solution:** The phrase "three times a number y" is represented by **3y** and "the sum of three times a number y and 7" means that you add 7 to that product. Thus, the algebraic expression is:
$$3y + 7$$
This expression indicates that you first multiply y by 3 and then add 7.

4. **Solution:** To solve the equation
$$4x - 5 = 11$$
follow these steps:

   (a) Add 5 to both sides to eliminate the constant on the left:
   $$4x - 5 + 5 = 11 + 5 \quad \implies \quad 4x = 16$$

(b) Divide both sides by 4 to isolate x:
$$\frac{4x}{4} = \frac{16}{4} \implies x = 4$$

Therefore, the solution to the equation is **x = 4**.

5. **Solution:** The equal sign in an equation is fundamental because it denotes that the expressions on both sides have the same value. In the example
$$2x + 3 = 7$$
the equal sign tells us that when the variable term **2x** and the constant **3** are combined, their sum is exactly the same as the number on the right side (**7**). This balance is central to the idea of an equation because any operation performed on one side must be correspondingly performed on the other to maintain equality. The equal sign, therefore, represents the idea of balance and equivalence in an equation.

6. **Solution:** In the expression
$$3x^2 - 4x + 8$$
the different parts are as follows:

- **Coefficients:** The number **3** is the coefficient for the term involving x^2 and **-4** is the coefficient for the term involving x. These numbers indicate how many times the variable parts are multiplied.
- **Variable Parts:** The term x^2 indicates that the variable x is raised to the power of 2, and the term x (which is x^1) is the variable by itself.
- **Constant:** The number **8** is the constant term because it does not multiply any variable and does not change.

Thus, the expression breaks down into:

- **$3x^2$**: Coefficient is 3, variable part is x^2.
- **$-4x$**: Coefficient is -4, variable part is x.
- **+8**: Constant term.

Each component plays its role in constructing a complete algebraic expression.

Chapter 2
Real Numbers and Their Properties

The Real Number System

1 Classification of Real Numbers

The real numbers form a complete set that includes several distinct categories based on the manner in which the numbers can be described and represented. The natural numbers are the basic counting numbers beginning from 1 and proceeding to infinity. Whole numbers extend the natural numbers by incorporating zero as an element. The integers encompass all whole numbers and their negatives, thereby providing a symmetric set about zero. Rational numbers are defined as those numbers that can be expressed in the form $\frac{a}{b}$, where a and b are integers and $b \neq 0$. This subset includes numbers that appear as fractions as well as decimals that terminate or repeat. Irrational numbers, in contrast, cannot be represented as a ratio of two integers; their decimal expansions are nonterminating and nonrepeating. Examples of such irrational numbers include quantities such as π and $\sqrt{2}$. The union of the rational and irrational numbers constitutes the set of real numbers, denoted by $\mathbb{R}$.

2 Representation on the Number Line

The real numbers are often depicted as points on a continuous line, known as the number line. This geometric representation highlights the ordered nature and the density of the real numbers. On the number line, every point corresponds to a unique real number, and the distance between any two points accurately reflects the difference in magnitude between the corresponding numbers. The continuity of the number line implies that between any two distinct real numbers a third real number can always be found. This property is particularly evident when considering fractions; between any two different fractions another fraction exists that lies strictly between them.

Properties of Arithmetic Operations

1 The Commutative Property

The commutative property applies to both addition and multiplication within the real number system. For addition, it is established that for any two real numbers a and b, the equality

$$a + b = b + a$$

holds true. Similarly, in multiplication, the relation

$$ab = ba$$

demonstrates that the order in which two numbers are multiplied does not affect the resulting product. This property is fundamental to the simplification of algebraic expressions and to the logical structure of arithmetic.

2 The Associative Property

The associative property concerns the grouping of numbers in operations and indicates that the manner in which numbers are grouped does not affect the final outcome of addition or multiplication. For three arbitrary real numbers a, b, and c, the associative property for addition is expressed as

$$(a + b) + c = a + (b + c).$$

In the context of multiplication, the relation

$$(ab)c = a(bc)$$

demonstrates a similar invariance under grouping. This property facilitates the reorganization of computational procedures by ensuring consistency regardless of the pairing of terms.

3 The Distributive Property

The distributive property provides a link between the operations of addition and multiplication by asserting that multiplication distributes over addition. For any real numbers a, b, and c, the distributive law is formulated as

$$a(b + c) = ab + ac.$$

This property is essential for expanding algebraic expressions and plays a critical role in the processes of factoring and simplification. It also allows the separation of complex expressions into simpler components that are easier to manipulate mathematically.

4 The Closure Property

The closure property ensures that the basic operations of addition, subtraction, and multiplication remain within the set of real numbers. Precisely, if a and b are elements of $\mathbb{R}$, then the results of the operations $a + b$, $a - b$, and ab are likewise contained in $\mathbb{R}$. This intrinsic property of the real number system guarantees the internal consistency required for the systematic development of algebraic theories and computations.

5 Identity and Inverse Properties

The real numbers are characterized by specific identity elements with respect to the operations of addition and multiplication. The additive identity, which is the number 0, satisfies the condition

$$a + 0 = a,$$

for any real number a. The multiplicative identity, represented by the number 1, fulfills the relation

$$a \cdot 1 = a.$$

Additionally, every real number possesses an additive inverse; for each a there exists a number $-a$ such that

$$a + (-a) = 0.$$

For every nonzero real number a, a multiplicative inverse also exists, denoted by $\frac{1}{a}$, which satisfies

$$a \cdot \frac{1}{a} = 1.$$

These properties form the backbone of algebraic manipulation and are indispensable in solving equations and in the study of more complex mathematical structures.

Multiple Choice Questions

1. Which of the following numbers is an irrational number?

    (a) 0.75

    (b) 0.6666...

    (c) $\sqrt{2}$

    (d) -5

2. Which property of the real numbers guarantees that the sum, difference, or product of two real numbers is also a real number?

    (a) Commutative Property

    (b) Associative Property

    (c) Closure Property

    (d) Distributive Property

3. The equation
$$a + b = b + a$$
    illustrates which property?

    (a) Associative Property

    (b) Commutative Property

    (c) Identity Property

    (d) Inverse Property

4. In the expression
$$a(b + c) = ab + ac,$$
    which property is being demonstrated?

    (a) Commutative Property

    (b) Associative Property

    (c) Distributive Property

    (d) Closure Property

5. What is the additive identity in the real number system?

    (a) 0

    (b) 1

    (c) -1

    (d) There is no additive identity.

6. Which statement best describes the continuity of the real numbers on the number line?

    (a) The number line has distinct, isolated points with gaps between them.

    (b) Every point on the number line corresponds to a unique real number, and between any two distinct real numbers there exists another real number.

    (c) Only rational numbers can be represented on the number line.

    (d) There are only finitely many points between any two given real numbers.

7. Which property indicates that the grouping of numbers in addition or multiplication does not affect the result?

    (a) Associative Property

    (b) Commutative Property

    (c) Identity Property

    (d) Inverse Property

Answers:

1. **C: $\sqrt{2}$**
   Explanation: $\sqrt{2}$ is an irrational number because its decimal expansion is nonterminating and nonrepeating, unlike 0.75 (a terminating decimal), 0.6666... (a repeating decimal), or the integer -5.

2. **C: Closure Property**
   Explanation: The closure property of the real numbers assures that when you add, subtract, or multiply any two real numbers, the result is also a real number.

3. **B: Commutative Property**
   Explanation: The equation $a+b = b+a$ shows that the order of addition does not change the sum; this is the commutative property of addition.

4. **C: Distributive Property**
   Explanation: The expression $a(b+c) = ab+ac$ demonstrates how multiplication distributes over addition, which is known as the distributive property.

5. **A: 0**
   Explanation: In the real number system, the additive identity is 0 because adding 0 to any number leaves it unchanged.

6. **B: Every point on the number line corresponds to a unique real number, and between any two distinct real numbers there exists another real number.**
   Explanation: The continuity of the real numbers means there are no gaps on the number line; between any two real numbers, you can always find another real number.

7. **A: Associative Property**
   Explanation: The associative property states that how numbers are grouped in an operation (either addition or multiplication) does not affect the result. For example, $(a+b)+c = a+(b+c)$.

Practice Problems

1. Classify the following numbers into the appropriate subsets of the real numbers (Natural numbers, Whole numbers, Integers, Rational numbers, and Irrational numbers):

$$7, \quad 0, \quad -4, \quad \frac{5}{3}, \quad 0.125, \quad \sqrt{2}$$

2. Plot and describe the positions on a number line for the following numbers. In your explanation, discuss the concept of density in the real number system:

$$-3, \quad 0, \quad \sqrt{2}, \quad 2.5$$

3. Simplify the expression using the distributive property:

$$5(2x + 3) - 3(2x + 3)$$

4. Given

$$a = -\frac{3}{4} \quad \text{and} \quad b = \frac{7}{2},$$

verify that the sum $a + b$ and the product ab are real numbers. Explain how this demonstrates the closure property for addition and multiplication.

5. Solve the equation:
$$5(x - 3) = 10,$$
and explain how the identity and inverse properties of real numbers are used in your solution.

6. Between the numbers
$$\frac{1}{2} \text{ and } \frac{3}{4},$$
find one rational number and one irrational number. Explain how these examples support the idea that the real number line is continuous.

Answers

1. **Solution:**
   We classify each number as follows:
   - **7**: This number is a Natural number (since natural numbers are generally defined as $1, 2, 3, \ldots$). Because every natural number is also a Whole number, an Integer, and can be expressed as $\frac{7}{1}$, it is a Rational number as well.

Hence, 7 is a Natural, Whole, Integer, Rational, and consequently a Real number.

- **0**: Zero is not typically considered a natural number, but it is a Whole number (which includes 0), an Integer, and a Rational number (since $0 = \frac{0}{1}$). Thus, 0 is a Real number.
- **−4**: The number −4 is an Integer (as it is a whole number with a negative sign) and a Rational number (since it can be written as $-4 = \frac{-4}{1}$). However, it is not a Natural or Whole number.
- **$\frac{5}{3}$**: This number is expressed as a ratio of two integers (with a nonzero denominator) and is therefore a Rational number. It is not an Integer, and by extension, not a Natural or Whole number.
- **0.125**: The decimal 0.125 terminates and can be written as a fraction ($0.125 = \frac{1}{8}$), making it a Rational number and, of course, a Real number.
- **$\sqrt{2}$**: The square root of 2 is a well-known Irrational number because its decimal expansion is nonterminating and nonrepeating. It cannot be expressed as the fraction of two integers, though it remains a Real number.

2. **Solution:**
On the number line:

- −3 is located to the left of 0.
- 0 is the central reference point (the origin).
- $\sqrt{2}$ is approximately 1.414 and lies to the right of 0, between 1 and 2.
- 2.5 is positioned further to the right, between 2 and 3.

The concept of density in the real number system means that between any two distinct real numbers, no matter how close together they are, there are infinitely many other real numbers. For example, between −3 and 0 or between $\sqrt{2}$ and 2.5, one can always find additional numbers. This property reflects the continuous nature of the number line.

3. **Solution:**
To simplify the expression:
$$5(2x + 3) - 3(2x + 3)$$

Notice that the term $(2x+3)$ is common to both products. Factor it out:
$$= (5-3)(2x+3)$$
Simplify the coefficient:
$$= 2(2x+3)$$
Finally, distribute the 2:
$$= 4x + 6$$
Thus, the simplified expression is $4x+6$.

4. **Solution:**
   We are given:
   $$a = -\frac{3}{4} \quad \text{and} \quad b = \frac{7}{2}$$
   First, compute the sum:
   $$a+b = -\frac{3}{4} + \frac{7}{2}$$
   To add these fractions, convert $\frac{7}{2}$ to have a denominator of 4:
   $$\frac{7}{2} = \frac{14}{4}$$
   Therefore:
   $$a+b = -\frac{3}{4} + \frac{14}{4} = \frac{11}{4}$$
   Next, compute the product:
   $$ab = \left(-\frac{3}{4}\right)\left(\frac{7}{2}\right) = -\frac{21}{8}$$
   Both $\frac{11}{4}$ and $-\frac{21}{8}$ are rational numbers, and by definition, all rational numbers are real. This confirms the closure property: performing addition or multiplication on two real numbers results in another real number.

5. **Solution:**
   We wish to solve:
   $$5(x-3) = 10$$

The first step is to isolate the term $(x-3)$. To do this, apply the multiplicative inverse of 5 (that is, divide both sides by 5):
$$x - 3 = \frac{10}{5} = 2$$
Next, use the additive inverse to solve for x by adding 3 to both sides:
$$x = 2 + 3 = 5$$
In this solution, the multiplicative inverse (dividing by 5) and the additive inverse (adding 3) are used to isolate the variable. These steps reflect the identity and inverse properties of real numbers, which ensure that the operations can be reversed to solve for unknowns.

6. **Solution:**
To find a rational number between $\frac{1}{2}$ and $\frac{3}{4}$, a simple method is to take the average:
$$\frac{\frac{1}{2} + \frac{3}{4}}{2} = \frac{\frac{2}{4} + \frac{3}{4}}{2} = \frac{\frac{5}{4}}{2} = \frac{5}{8}$$
The number $\frac{5}{8}$ is a rational number.

For an irrational number, note that adding a nonzero irrational number to a rational number results in an irrational number. For instance, consider:
$$\frac{5}{8} + \frac{\sqrt{2}}{100}$$
Here, $\frac{5}{8} \approx 0.625$ and $\frac{\sqrt{2}}{100} \approx 0.0141$, so their sum is approximately 0.6391. Because $\frac{\sqrt{2}}{100}$ is irrational, the resulting number is also irrational.

These examples demonstrate that between any two distinct rational numbers there exist both rational and irrational numbers. This fact underlines the continuity (and density) of the real number line, meaning there are no gaps between numbers.

Chapter 3

Order of Operations and Expression Evaluation

The Hierarchy of Operations

1 Parentheses and Grouping Symbols

Expressions are first simplified by identifying and resolving any computations enclosed within grouping symbols. Parentheses, brackets, and braces indicate parts of an expression that must be evaluated as a single entity before any other operations are performed. The process adheres to a systematic approach in which nested grouping symbols are processed from the innermost set outward, thereby ensuring that the intended structure of the expression is maintained and ambiguity is eliminated from the evaluation process.

2 Exponents and Root Operations

Following the resolution of grouped expressions, exponentiation, including roots and powers, is addressed. Exponential operations represent repeated multiplication and serve as a concise way to denote large or fractional products. In addition to power functions, radical expressions, which denote the extraction of roots, are situated in the same level of precedence as exponents. This placement within the hierarchy ensures that all instances of repeated multiplications or divisions by means of roots receive priority in the overall sequence of operations.

3 Multiplication and Division

Operations of multiplication and division follow closely in the sequence of evaluation. These operations hold equal precedence and are executed in a left-to-right order. When multiple multiplication and division operations appear consecutively within an expression, the evaluation proceeds sequentially from the leftmost operation to the rightmost. This uniform approach avoids misinterpretation and preserves the intended relationships among the factors and divisors within complex expressions.

4 Addition and Subtraction

The final stage in the operation hierarchy encompasses addition and subtraction. Like multiplication and division, addition and subtraction share the same level of precedence and are carried out from left to right. The methodical execution of these operations reduces the expression gradually to a simplified numerical or algebraic form. The left-to-right processing ensures that each subsequent operation builds logically on the outcome of previous computations, thereby maintaining consistency throughout the evaluation.

Evaluation of Algebraic Expressions

The rules of operation precedence extend naturally to expressions containing variables as well as numerical constants. In algebraic expressions, it is essential to treat each component—whether a literal or a number—in accordance with the established hierarchy. Parentheses serve to group variable terms and constants alike, allowing for the isolation of specific segments of the expression that require independent evaluation. Once values or variable expressions are substituted into the structure, the same sequential process of addressing grouping symbols, exponents, multiplication or division, and finally addition or subtraction is observed. This systematic approach not only yields correct numerical values but also simplifies the combination of like terms and other algebraic manipulations.

Strategies for Methodical Evaluation

Expressions that incorporate several layers of operations require careful sequential reduction. The initial step involves scanning the entire expression to locate grouping symbols and nested operations. Once the innermost expressions are simplified, the evaluation progresses by applying the exponentiation directive. Thereafter, sequential multiplication and division steps follow, processing operations from left to right before proceeding to the final phase of addition and subtraction. This structured method minimizes errors that may arise from misinterpretation of the expression's arrangement. Ambiguous cases, particularly those involving operations of equal precedence, are resolved by adhering strictly to the left-to-right convention, thereby preserving the integrity of the evaluation process. Each transformation in the expression is performed with an emphasis on maintaining consistency with the fundamental arithmetic laws, ensuring that the resultant form accurately mirrors the original structure and intent of the algebraic expression.

Multiple Choice Questions

1. What is the first step in evaluating an algebraic expression according to the order of operations?

   (a) Perform addition and subtraction.
   (b) Evaluate multiplication and division.
   (c) Simplify expressions inside grouping symbols (parentheses, brackets, braces).
   (d) Apply exponentiation and radical operations.

2. When an expression contains nested grouping symbols, which grouping should be simplified first?

   (a) The outermost grouping symbols.
   (b) The grouping symbols with the largest numbers.
   (c) All grouping symbols simultaneously.
   (d) The innermost grouping symbols.

3. After resolving grouping symbols, which set of operations is performed next?

(a) Addition and subtraction.

   (b) Multiplication and division.

   (c) Exponentiation and radical operations.

   (d) Subtraction and multiplication.

4. When multiple multiplication and division operations appear consecutively in an expression, what is the proper method to evaluate them?

   (a) Perform all multiplication first, then all division.

   (b) Perform all division first, then all multiplication.

   (c) Evaluate them simultaneously.

   (d) Process the operations sequentially from left to right.

5. In an expression that contains both addition and subtraction, how should these operations be executed?

   (a) All addition operations must be completed before any subtraction.

   (b) All subtraction operations must be completed before any addition.

   (c) They are executed from right to left.

   (d) They are processed sequentially from left to right.

6. After substituting numerical values for variables in an algebraic expression, what must be done to ensure accuracy in evaluation?

   (a) The order of operations can be disregarded once variables are replaced.

   (b) The simplified expression can be computed in any random order.

   (c) The computation must follow the established order: grouping symbols, exponents (and radicals), multiplication and division (left to right), then addition and subtraction (left to right).

   (d) Only the arithmetic operations (addition and subtraction) need to be followed.

7. What is the primary benefit of using a methodical, step-by-step evaluation strategy for complex expressions?

(a) It allows skipping steps to reach a quicker answer.

(b) It ensures a consistent, error-free process that honors arithmetic laws.

(c) It makes the algebraic expression more challenging.

(d) It permits the operations to be executed in any order.

Answers:

1. **C: Simplify expressions inside grouping symbols (parentheses, brackets, braces)**
   According to the order of operations, grouping symbols are the first to be addressed because they isolate parts of an expression that must be evaluated as a unit.

2. **D: The innermost grouping symbols**
   In nested grouping symbols, simplifying the innermost group first removes ambiguity and ensures that each step respects the structural hierarchy of the expression.

3. **C: Exponentiation and radical operations**
   Once grouping symbols are resolved, exponentiation (including roots) is performed next as it represents repeated multiplication and the extraction of roots.

4. **D: Process the operations sequentially from left to right**
   Multiplication and division have equal precedence; hence, when they appear together, they are handled in the order they appear—from left to right—to maintain accuracy.

5. **D: They are processed sequentially from left to right**
   Like multiplication and division, addition and subtraction share equal precedence and must be executed in the order they occur from left to right.

6. **C: The computation must follow the established order: grouping symbols, exponents (and radicals), multiplication and division (left to right), then addition and subtraction (left to right)**
   Even after substituting variables with numerical values, the structure of the expression remains unchanged; therefore, the standard order of operations must be followed to ensure correct evaluation.

7. **B: It ensures a consistent, error-free process that honors arithmetic laws**
   A methodical, step-by-step evaluation minimizes errors and preserves the intended structure and meaning of an expression by adhering to the fundamental arithmetic rules.

Practice Problems

1. Evaluate the following expression by applying the order of operations:
$$7 + 3 \cdot (4-2)^2 - \frac{8}{2}$$

2. Evaluate the following expression with nested grouping symbols:
$$2\left(3 + (4-1)^2\right) - 5$$

3. Evaluate the following expression involving square roots and exponents:
$$\sqrt{9+7-4} + \frac{2^3}{4}$$

4. Substitute the value $x = -1$ into the expression and simplify:
$$3x^2 - 2x + 5$$

5. Simplify the following expression using the order of operations:
$$\frac{12 + 8}{4 - 2} + 7 \cdot 1^2$$

6. Evaluate the expression step-by-step:
$$(2 + 3 \cdot (1 + 2))^2 - 4 \cdot \sqrt{16}$$

Answers

1. **Solution:**
   Begin by evaluating the expression inside the parentheses:
   $$4 - 2 = 2.$$
   Next, apply the exponent to the result:
   $$(2)^2 = 4.$$
   Then perform the multiplication and division:
   $$3 \cdot 4 = 12 \quad \text{and} \quad \frac{8}{2} = 4.$$
   Finally, combine these results with the addition and subtraction:
   $$7 + 12 - 4 = 15.$$
   Therefore, the expression simplifies to
   $$15$$
   .

2. **Solution:**
   Start with the innermost grouping:
   $$4 - 1 = 3.$$
   Then, apply the exponent to this result:
   $$(3)^2 = 9.$$

Substitute back into the expression inside the larger grouping:
$$3 + 9 = 12.$$

Multiply by 2:
$$2 \cdot 12 = 24.$$

Finally, subtract 5:
$$24 - 5 = 19.$$

Thus, the simplified expression equals
$$19$$
.

3. **Solution:**
First, simplify the expression inside the square root:
$$9 + 7 - 4 = 12.$$

Then, take the square root of the result:
$$\sqrt{12} = 2\sqrt{3} \quad (\text{since } 12 = 4 \times 3).$$

Next, evaluate the exponent and division:
$$2^3 = 8 \quad \text{and} \quad \frac{8}{4} = 2.$$

Finally, add the two parts:
$$2\sqrt{3} + 2.$$

Therefore, the final result is
$$2\sqrt{3} + 2$$
.

4. **Solution:**
Substitute $x = -1$ into the expression:
$$3(-1)^2 - 2(-1) + 5.$$

Evaluate the exponent:
$$(-1)^2 = 1,$$

so the expression becomes:
$$3(1) + 2 + 5.$$
Combine the terms:
$$3 + 2 + 5 = 10.$$
Therefore, when $x = -1$, the expression evaluates to
$$10$$

5. **Solution:**
   Begin by computing the numerator:
   $$12 + 8 = 20.$$
   Then compute the denominator:
   $$4 - 2 = 2.$$
   Divide to obtain:
   $$\frac{20}{2} = 10.$$
   Next, evaluate the exponent:
   $$1^2 = 1,$$
   and multiply:
   $$7 \cdot 1 = 7.$$
   Finally, add the two results:
   $$10 + 7 = 17.$$
   Thus, the simplified expression is
   $$17$$

6. **Solution:**
   Start by evaluating the innermost parentheses:
   $$1 + 2 = 3.$$

Multiply by 3:
$$3 \cdot 3 = 9.$$

Add 2:
$$2 + 9 = 11.$$

Square the result:
$$11^2 = 121.$$

Separately, evaluate the square root:
$$\sqrt{16} = 4.$$

Multiply by 4:
$$4 \cdot 4 = 16.$$

Lastly, subtract the product from the squared term:
$$121 - 16 = 105.$$

Therefore, the evaluated expression equals
$$105$$

.

Chapter 4

Combining Like Terms and the Distributive Property

Identifying Like Terms

In algebraic expressions, a term is defined as a single number, a variable, or the product of numbers and variables. Terms that contain identical variable factors raised to the same exponents are classified as like terms. For instance, in the expression

$$7x + 3 - 2x + 5,$$

the terms $7x$ and $-2x$ are like terms because each has the variable x to the first power, while the constants 3 and 5 are also like terms by virtue of both being numerical constants. The process of identifying like terms involves isolating the variable part of each term to determine whether different terms may be combined in a meaningful way.

Combining Like Terms in Algebraic Expressions

Combining like terms involves the consolidation of terms that share the same variable components into a single term by performing the appropriate arithmetic operations on their coefficients. This technique is rooted in the commutative and associative properties of addition. Consider the expression

$$3x + 4 + 5x - 2.$$

The like terms $3x$ and $5x$ are combined by adding their coefficients, yielding
$$(3 + 5)x = 8x.$$

Simultaneously, the constant terms 4 and -2 are combined to produce 2. Consequently, the simplified form of the expression is

$$8x + 2.$$

This method of reduction not only simplifies the appearance of the expression but also lays the groundwork for solving equations and further algebraic manipulations.

The Distributive Property and Its Applications

The distributive property describes the process by which multiplication is distributed over addition or subtraction within parentheses. Mathematically, the property is stated as

$$a(b + c) = ab + ac,$$

where a, b, and c represent constants or expressions. This property is applicable irrespective of whether the operation within the parentheses is addition or subtraction. An elementary example is seen in the expression
$$4(x + 3),$$
which expands to
$$4x + 12.$$

The expansion enables the expression to be manipulated further by isolating like terms or facilitating the solution of an equation. The distributive property is essential when an expression contains factors that are applied to sums or differences, ensuring that each component within the grouping symbol is appropriately scaled by the multiplying factor.

1 Intertwining Distribution with Combination of Like Terms

In practice, many algebraic expressions require both the expansion provided by the distributive property and the subsequent combination of like terms to achieve a simplified form. Consider the compound expression

$$3(2y + 4) - 5y.$$

Initially, the distributive property is applied to eliminate the parentheses:
$$3 \cdot 2y + 3 \cdot 4 - 5y = 6y + 12 - 5y.$$

Subsequently, the like terms $6y$ and $-5y$ are combined to yield

$$(6 - 5)y = y.$$

Thus, the expression simplifies to

$$y + 12.$$

This layered approach—first distributing and then combining like terms—establishes a clear pathway for reducing complex expressions into simpler and more manageable forms, facilitating further manipulation or evaluation in algebraic processes.

Multiple Choice Questions

1. Which pair of terms in the expression
   $$5x^2 + 3x - 2x^2 + 7$$
   are considered like terms?

   (a) $5x^2$ and $3x$

   (b) $5x^2$ and $-2x^2$

   (c) $3x$ and 7

   (d) $-2x^2$ and 7

2. What is the simplified form of the expression
   $$3(2y + 5) - 4y?$$

(a) $6y + 5 - 4y$

(b) $2y + 15$

(c) $2y + 5$

(d) $-2y + 15$

3. Which property justifies rewriting
$$4(x + 3)$$
as
$$4x + 12?$$

   (a) Commutative property

   (b) Associative property

   (c) Distributive property

   (d) Identity property

4. Simplify the expression
$$7x + 3 - 2x + 5.$$

   (a) $5x + 8$

   (b) $9x + 8$

   (c) $5x + 2$

   (d) $7x + 8$

5. Which of the following best defines "like terms" in an algebraic expression?

   (a) Terms that always have the same sign.

   (b) Terms that have the same numerical coefficient.

   (c) Terms that contain identical variable parts raised to the same exponent.

   (d) Terms that can be multiplied directly.

6. After using the distributive property and combining like terms, what is the simplified form of
$$2(3x - 4) + 4(x + 2)?$$

   (a) $10x$

(b) $10x + 2$
(c) $8x$
(d) $6x$

7. Which of the following is an incorrect use of the distributive property?

    (a) $3(2x + 5) = 6x + 15$
    (b) $-2(3x - 4) = -6x + 8$
    (c) $4(x + 0) = 4x$
    (d) $5(x - 2) = 5x - 10x$

Answers:

1. **B: $5x^2$ and $-2x^2$**
   These terms are like terms because they both contain the variable x raised to the second power. The coefficients (5 and -2) can be combined.

2. **B: $2y + 15$**
   First, distribute: $3(2y + 5) = 6y + 15$. Then, subtract $4y$ to combine like terms: $6y - 4y + 15 = 2y + 15$.

3. **C: Distributive property**
   The transformation from $4(x + 3)$ to $4x + 12$ is performed by applying the distributive property, which states that $a(b + c) = ab + ac$.

4. **A: $5x + 8$**
   Combine the like terms: $7x - 2x = 5x$ and $3 + 5 = 8$, so the simplified expression is $5x + 8$.

5. **C: Terms that contain identical variable parts raised to the same exponent**
   Like terms must have the exact same variable factors and exponents. The numerical coefficients may differ, but the variable component must be identical for terms to be combined.

6. **A: $10x$**
   Distribute each term:
   $$2(3x - 4) = 6x - 8 \quad \text{and} \quad 4(x + 2) = 4x + 8.$$
   Then combine like terms: $6x + 4x = 10x$ and $-8 + 8 = 0$, leaving $10x$.

7. **D:** $5(x-2) = 5x - 10x$

   The correct application of the distributive property in this case should yield $5(x-2) = 5x - 10$. Option D incorrectly multiplies -2 by 5, resulting in an erroneous $-10x$ rather than -10.

Practice Problems

1. Simplify the following expression by combining like terms:

$$5x + 3 - 2x + 7 + 4y - 3y$$

2. Simplify the following expression by using the distributive property and then combining like terms:

$$3(2x + 5) - 4(x - 2)$$

3. Simplify the expression by first applying the distributive property:

$$2(3a + 4) + 3(2a - 1) - 5$$

4. Simplify the expression:
$$4(x+2) - 3(2x-5) + 2(3-x)$$

5. Simplify the following expression using the distributive property and then combining like terms:
$$3(2y-4) + 2(3y+6) - 5y$$

6. Simplify the expression by applying the distributive property and combining like terms:
$$2(3m + 4) - m(5 - 2m) + 3$$

Answers

1. For the expression:
$$5x + 3 - 2x + 7 + 4y - 3y$$

   Solution:
   Identify and group the like terms. The terms with the variable x are:
   $$5x - 2x = 3x.$$
   The terms with the variable y are:
   $$4y - 3y = y.$$
   The constant terms are:
   $$3 + 7 = 10.$$
   Thus, the simplified expression is:
   $$3x + y + 10.$$

2. For the expression:
$$3(2x + 5) - 4(x - 2)$$

Solution:
First, use the distributive property on each grouped term:
$$3(2x + 5) = 6x + 15,$$
$$-4(x - 2) = -4x + 8.$$
Now, combine the like terms. The x-terms are:
$$6x - 4x = 2x,$$
and the constant terms are:
$$15 + 8 = 23.$$
Therefore, the simplified expression is:
$$2x + 23.$$

3. For the expression:
$$2(3a + 4) + 3(2a - 1) - 5$$

Solution:
Begin by distributing the multiplication over addition:
$$2(3a + 4) = 6a + 8,$$
$$3(2a - 1) = 6a - 3.$$
Now, add the results and subtract 5:
$$6a + 8 + 6a - 3 - 5.$$
Combine the like terms. The a-terms add up to:
$$6a + 6a = 12a,$$
and the constants simplify as:
$$8 - 3 - 5 = 0.$$
Thus, the final simplified expression is:
$$12a.$$

4. For the expression:
$$4(x+2) - 3(2x-5) + 2(3-x)$$

Solution:
First, apply the distributive property to eliminate the parentheses:
$$4(x+2) = 4x + 8,$$
$$-3(2x-5) = -6x + 15,$$
$$2(3-x) = 6 - 2x.$$

Next, combine like terms. Group the x-terms:
$$4x - 6x - 2x = -4x,$$

and then add the constants:
$$8 + 15 + 6 = 29.$$

Therefore, the expression simplifies to:
$$-4x + 29.$$

5. For the expression:
$$3(2y-4) + 2(3y+6) - 5y$$

Solution:
Begin with the distributive property:
$$3(2y-4) = 6y - 12,$$
$$2(3y+6) = 6y + 12.$$

Now, including the $-5y$ term:
$$6y - 12 + 6y + 12 - 5y.$$

Combine the y-terms:
$$6y + 6y - 5y = 7y,$$

and note that the constants cancel:
$$-12 + 12 = 0.$$

So, the simplified expression is:
$$7y.$$

6. For the expression:
$$2(3m + 4) - m(5 - 2m) + 3$$

Solution:
First, apply the distributive property:
$$2(3m + 4) = 6m + 8.$$

Next, expand the term with m:
$$m(5 - 2m) = 5m - 2m^2.$$

Since this term is subtracted, it becomes:
$$-5m + 2m^2.$$

Now, add the constant term:
$$6m + 8 - 5m + 2m^2 + 3.$$

Combine like terms by grouping the m^2-term, the m-terms, and the constants:
$$2m^2 \quad (m^2\text{-term}),$$
$$6m - 5m = m \quad (m\text{-term}),$$
$$8 + 3 = 11 \quad (\text{constant}).$$

Thus, the expression simplifies to:
$$2m^2 + m + 11.$$

Chapter 5

Exponents and Powers: Concepts and Applications

Definition and Notation

Exponents are a method of expressing repeated multiplication in a compact form. An expression of the form a^n indicates that the base a is multiplied by itself n times. Here, a represents any real number or algebraic expression, and n is a nonnegative integer that serves as the count of multiplications. For example, writing a^4 is equivalent to
$$a \cdot a \cdot a \cdot a.$$
This notation is essential in algebra as it not only simplifies the representation of large products but also lays the groundwork for developing systematic methods to manipulate and simplify expressions.

Multiplication of Exponential Terms

When multiplying exponential terms that share the same base, the exponents are added according to the rule
$$a^m \cdot a^n = a^{m+n}.$$

This property stems from the definition of exponents as repeated multiplication. For instance, consider the product $x^2 \cdot x^3$. By interpreting the expression as

$$(x \cdot x) \cdot (x \cdot x \cdot x),$$

it becomes apparent that the base x is multiplied a total of $2 + 3$ times, leading to the simplified result

$$x^{2+3} \quad \text{or} \quad x^5.$$

This method of combining exponents enhances the ability to condense algebraic expressions into more manageable forms.

Division of Exponential Terms

Division involving exponential expressions with the same base is governed by the rule

$$\frac{a^m}{a^n} = a^{m-n},$$

provided that the base a is nonzero. This rule is derived from inverting the product formulation for exponents. Taking an expression such as

$$\frac{y^7}{y^4}$$

as an example, the exponent in the numerator exceeds that in the denominator by $7 - 4$, which reduces the expression to

$$y^3.$$

The process of subtracting the exponent in the denominator from the exponent in the numerator simplifies rational expressions and assists in carrying out further algebraic manipulations.

The Power of a Power Rule

In scenarios where an exponential expression is itself raised to another power, the exponents multiply according to the rule

$$(a^m)^n = a^{m \cdot n}.$$

This property follows directly from repeated application of exponentiation. For example, the expression $(z^3)^2$ calls for multiplying the exponents 3 and 2, ultimately yielding

$$z^{3 \cdot 2} \quad \text{or} \quad z^6.$$

Employing the power of a power rule simplifies complex exponentiation tasks and is instrumental when dealing with nested exponents, reducing them to an equivalent single exponent expression.

Application to Algebraic Expressions

The laws governing exponents play a crucial role in the manipulation and simplification of algebraic expressions. By employing the rule for multiplication of exponents, expressions involving multiple occurrences of a given base can be efficiently condensed. Similarly, the division rule allows for the reduction of expressions where a common base appears in both the numerator and denominator, transforming potentially unwieldy fractions into simplified power forms. The power of a power rule is equally valuable when faced with expressions that involve multiple layers of exponentiation. Mastery of these laws not only aids in the simplification process but also facilitates the resolution of equations and the evaluation of polynomial expressions, thereby strengthening overall algebraic proficiency.

Multiple Choice Questions

1. Which of the following best describes the meaning of the notation a^n (with n a positive integer)?

    (a) a added to itself n times

    (b) a multiplied by n

    (c) a multiplied by itself n times

    (d) a raised to the power of n squared

2. When multiplying exponential terms with the same base, such as $a^m \cdot a^n$, which rule applies?

    (a) $a^m \cdot a^n = a^{m \cdot n}$

    (b) $a^m \cdot a^n = a^{m+n}$

(c) $a^m \cdot a^n = a^{m-n}$

(d) $a^m \cdot a^n = a^{n-m}$

3. How would you simplify the expression $(b^6)/(b^2)$, assuming b 0?

   (a) b^4
   (b) b^8
   (c) b^{12}
   (d) b^3

4. Which expression correctly applies the power of a power rule to $(x^3)^4$?

   (a) x^7
   (b) x^{12}
   (c) $x^{1/12}$
   (d) x^{3+4}

5. Simplify the following expression: $(2^3 \cdot 2^4) / 2^5$.

   (a) 2^2
   (b) 2^6
   (c) 2^3
   (d) 2^5

6. Which property of exponents is illustrated by rewriting $(5^3)^2$ as 5^6?

   (a) The product of powers rule
   (b) The quotient of powers rule
   (c) The power of a power rule
   (d) The zero exponent rule

7. Consider the expression $(x^2 \cdot x^5) / x^4$. What is its simplified form?

   (a) x^3
   (b) x^7
   (c) x^8
   (d) x

Answers:

1. **C: a multiplied by itself n times** Explanation: The notation a^n means that the base a is used in multiplication n times. For example, a^4 represents a · a · a · a.

2. **B: $a^m \cdot a^n = a^{m+n}$** Explanation: When multiplying exponential terms with the same base, you add their exponents. This rule directly stems from the definition of exponents as repeated multiplication.

3. **A: b^4** Explanation: Dividing like bases means subtracting the exponents: $b^6 / b^2 = b^{6-2} = b^4$.

4. **B: x^{12}** Explanation: The power of a power rule states that $(x^m)^n = x^{m \cdot n}$. Thus, $(x^3)^4 = x^{3 \times 4} = x^{12}$.

5. **A: 2^2** Explanation: First, apply the multiplication rule: $2^3 \cdot 2^4 = 2^{3+4} = 2^7$. Then, applying the division rule, $2^7 / 2^5 = 2^{7-5} = 2^2$.

6. **C: The power of a power rule** Explanation: The operation $(5^3)^2$ uses the power of a power rule, which requires multiplying the exponents: $3 \times 2 = 6$, so the expression simplifies to 5^6.

7. **A: x^3** Explanation: Multiply the powers first: $x^2 \cdot x^5 = x^{2+5} = x^7$. Then, applying the division rule, $x^7 / x^4 = x^{7-4} = x^3$.

Practice Problems

1. Simplify the following expression using the multiplication rule for exponents:
$$x^3 \cdot x^5$$

2. Simplify the following expression using the division rule for exponents:
$$\frac{y^8}{y^3}$$

3. Simplify the following expression using the power of a power rule:
$$(z^2)^4$$

4. Simplify the following expression, carefully applying the exponent rules to both the coefficient and the variable:
$$\frac{(2x^3)^2}{2x}$$

5. Simplify the following expression by applying the exponent rules:
$$\frac{(3a^2b)^3}{9a^4b^2}$$

6. Simplify the following expression by first simplifying inside the parentheses and then applying the power rule:
$$\left(\frac{x^5y^3}{x^2y}\right)^2$$

Answers

1. **Solution:**
   To simplify
   $$x^3 \cdot x^5,$$
   we use the multiplication rule for exponents which states that when multiplying expressions with the same base, we add the exponents:
   $$x^3 \cdot x^5 = x^{3+5} = x^8.$$

Therefore, the simplified result is
$$x^8.$$

2. **Solution:**
   For the expression
   $$\frac{y^8}{y^3},$$
   apply the division rule for exponents which tells us to subtract the exponent in the denominator from the exponent in the numerator:
   $$\frac{y^8}{y^3} = y^{8-3} = y^5.$$
   Thus, the simplified expression is
   $$y^5.$$

3. **Solution:**
   The expression
   $$(z^2)^4$$
   is simplified using the power of a power rule. This rule states that when an exponent is raised to another exponent, you multiply the exponents:
   $$(z^2)^4 = z^{2\cdot 4} = z^8.$$
   So, the result is
   $$z^8.$$

4. **Solution:**
   First, simplify the numerator $(2x^3)^2$ by applying the power of a product rule:
   $$(2x^3)^2 = 2^2 \cdot (x^3)^2 = 4x^{3\cdot 2} = 4x^6.$$
   Now, the expression becomes:
   $$\frac{4x^6}{2x}.$$
   Next, simplify the coefficient and the variables separately. For the coefficients:
   $$\frac{4}{2} = 2.$$

For the variables, use the division rule:
$$\frac{x^6}{x} = x^{6-1} = x^5.$$

Combining these results:
$$\frac{4x^6}{2x} = 2x^5.$$

Therefore, the simplified expression is
$$2x^5.$$

5. **Solution:**
   Begin by expanding the numerator:
   $$(3a^2b)^3 = 3^3 \cdot (a^2)^3 \cdot b^3 = 27a^{2\cdot 3}b^3 = 27a^6b^3.$$

   The denominator remains as:
   $$9a^4b^2.$$

   Now, form the fraction:
   $$\frac{27a^6b^3}{9a^4b^2}.$$

   Simplify the coefficients:
   $$\frac{27}{9} = 3.$$

   Apply the division rule for the variables:
   $$\frac{a^6}{a^4} = a^{6-4} = a^2, \quad \frac{b^3}{b^2} = b^{3-2} = b.$$

   Combine these results:
   $$\frac{27a^6b^3}{9a^4b^2} = 3a^2b.$$

   So, the expression simplifies to
   $$3a^2b.$$

6. **Solution:**
Start by simplifying the expression inside the parentheses:
$$\frac{x^5 y^3}{x^2 y} = x^{5-2} \cdot y^{3-1} = x^3 y^2.$$

Next, raise the simplified expression to the power of 2 using the power of a power rule:
$$(x^3 y^2)^2 = x^{3 \cdot 2} \cdot y^{2 \cdot 2} = x^6 y^4.$$

Therefore, the final simplified expression is
$$x^6 y^4.$$

Chapter 6

Scientific Notation and Special Products

Scientific Notation

Scientific notation provides a systematic method for representing very large numbers or very small numbers in a compact and efficient form. In this notation, a number is expressed as the product of a decimal coefficient and a power of 10. The coefficient is chosen to be a number that satisfies the inequality 1 a < 10, and it is multiplied by 10 raised to an integer exponent. In mathematical terms, any number N can be written in the form

$$N = a \times 10^n,$$

where a is the significant figure and n indicates the number of places the decimal point has been shifted. For instance, when converting a large number such as 300000 into scientific notation, the coefficient becomes 3.0 and the decimal is shifted five places to the left, yielding

$$3.0 \times 10^5.$$

Similarly, a very small number like 0.00045 is expressed by shifting the decimal point to produce a coefficient of 4.5, with the decimal movement recorded by a negative exponent,

$$4.5 \times 10^{-4}.$$

The conversion process involves identifying the significant digits and determining the exact number of positions required to reestablish the standard form of the coefficient. The power of 10 then denotes the magnitude and the scale of the original number, simplifying both computation and interpretation in further algebraic operations.

Special Products

Special products are algebraic identities that allow for rapid expansion of expressions without resorting to repetitive application of the distributive property. Recognizing these formulas is an essential skill in algebra, as they reveal structural patterns and significantly reduce computational effort. Two of the most frequently encountered identities are those involving the square of a binomial and the product of a sum and a difference.

1 Square of a Binomial

The square of a binomial is an identity that comes in two forms, depending on whether the binomial involves addition or subtraction. For an expression in the form of

$$(a+b)^2,$$

the expansion is performed by multiplying the binomial by itself, which results in

$$(a+b)^2 = a^2 + 2ab + b^2.$$

Each term in the binomial is multiplied with every other term, and like terms are combined to yield a perfect square trinomial. In the case of a binomial with subtraction,

$$(a-b)^2,$$

the expansion appropriately accounts for the negative sign associated with the second term, yielding

$$(a-b)^2 = a^2 - 2ab + b^2.$$

The symmetry in these identities simplifies many algebraic problems, as the square of a binomial formula is often used to quickly evaluate or factor expressions without executing full distributive multiplication.

2 Product of a Sum and a Difference

Another valuable identity in the realm of special products is the product of a sum and a difference, which is given by

$$(a+b)(a-b) = a^2 - b^2.$$

In this case, the cross terms that would normally appear in the expansion cancel out. The result is a difference of two squares, which encapsulates the expression in a much simpler form. This identity is particularly useful when factoring expressions or when simplifying equations, as it provides an immediate path to reducing the product of two binomials into a subtraction of squared terms.

Scientific notation and these special product formulas exemplify techniques that enhance both the clarity and computational efficiency in algebra. The methodical nature of scientific notation facilitates the handling of numbers across extreme scales, while the recognized patterns of special products lead to rapid expansion and simplification of polynomial expressions.

Multiple Choice Questions

1. Which of the following describes the correct range for the coefficient in scientific notation?

    (a) $0 \leq a < 1$

    (b) $0 < a \leq 10$

    (c) $1 \leq a < 10$

    (d) a can be any real number

2. Which of the following is the correct scientific notation for 300000?

    (a) 3.0×10^5

    (b) 3.0×10^{-5}

    (c) 30×10^4

    (d) 0.3×10^6

3. Which of the following correctly represents 0.00045 in scientific notation?

    (a) 0.45×10^{-3}

(b) 4.5×10^{-4}

(c) 45×10^{-5}

(d) 4.5×10^4

4. What does the exponent in a number written in scientific notation indicate?

    (a) The number of positions the decimal point has been shifted.
    (b) The number of significant figures in the coefficient.
    (c) The number of zeros in the original number.
    (d) The error margin in the approximation.

5. Which of the following is the correct expansion of the binomial $(a+b)^2$?

    (a) $a^2 + b^2$
    (b) $a^2 + 2ab + b^2$
    (c) $a^2 - 2ab + b^2$
    (d) $2a^2 + 2ab$

6. What is the simplified result of multiplying $(a+b)$ and $(a-b)$?

    (a) $a^2 + b^2$
    (b) $a^2 - 2ab + b^2$
    (c) $a^2 - b^2$
    (d) $2ab$

7. What is a primary benefit of recognizing and using special product formulas in algebra?

    (a) They simplify expansion and factoring by reducing repetitive multiplication.
    (b) They eliminate the need for the distributive property entirely.
    (c) They allow expressions to be solved without isolating variables.
    (d) They increase the number of terms, improving clarity.

Answers:

1. **C:** $1 \leq a < 10$
   In scientific notation, the coefficient must be chosen so that it is at least 1 but less than 10. This standardizes the representation of numbers and simplifies comparison of magnitudes.

2. **A:** 3.0×10^5
   To express 300000 in scientific notation, the decimal point is moved 5 places to the left (yielding 3.0), resulting in the form 3.0×10^5.

3. **B:** 4.5×10^{-4}
   For 0.00045, the decimal is shifted 4 places to the right to obtain the coefficient 4.5. Since the number is less than 1, the exponent is negative, giving 4.5×10^{-4}.

4. **A: The number of positions the decimal point has been shifted**
   The exponent in scientific notation indicates how many positions the decimal point has been moved from its original position; a positive exponent indicates a shift to the left (for large numbers) and a negative exponent indicates a shift to the right (for small numbers).

5. **B:** $a^2 + 2ab + b^2$
   The square of a binomial, $(a+b)^2$, expands by multiplying the binomial by itself, resulting in $a^2 + 2ab + b^2$. This is a standard identity used frequently for quick expansion.

6. **C:** $a^2 - b^2$
   Multiplying $(a+b)(a-b)$ uses the difference of squares identity, which cancels the middle terms to yield $a^2 - b^2$.

7. **A: They simplify expansion and factoring by reducing repetitive multiplication.**
   Special product formulas, such as the square of a binomial or the difference of squares, help streamline algebraic manipulations by reducing the number of steps required in expansion and factoring, thereby saving time and minimizing errors.

Practice Problems

1. Convert the following large number into scientific notation:

$$3600000$$

2. Convert the following small number into scientific notation:

$$0.00052$$

3. Expand the following binomial using the square of a sum identity:

$$(4x+7)^2$$

4. Expand the following binomial using the square of a difference identity:

$$(5-2y)^2$$

5. Expand the following expression using the product of a sum and a difference identity:
$$(6x + 8)(6x - 8)$$

6. Multiply the following numbers in scientific notation and express the answer in scientific notation:
$$(3.0 \times 10^5) \times (2.0 \times 10^3)$$

Answers

1. **Solution:** To convert 3600000 into scientific notation, rewrite it so that the coefficient is between 1 and 10. Move the decimal point 6 places to the left:
$$3600000 = 3.6 \times 10^6.$$
Here, 3.6 is the significant figure and 10 raised to the 6 indicates the decimal was shifted 6 places.

2. **Solution:** To write 0.00052 in scientific notation, shift the decimal point to the right until you have a coefficient between 1 and 10. Moving the decimal 4 places to the right gives:
$$0.00052 = 5.2 \times 10^{-4}.$$
The negative exponent indicates that the decimal point was moved to the right.

3. **Solution:** To expand $(4x+7)^2$, use the square of a binomial formula:
$$(a+b)^2 = a^2 + 2ab + b^2.$$
Let $a = 4x$ and $b = 7$. Then,
$$(4x)^2 = 16x^2, \quad 2(4x)(7) = 56x, \quad \text{and} \quad 7^2 = 49.$$
Therefore,
$$(4x+7)^2 = 16x^2 + 56x + 49.$$

4. **Solution:** To expand $(5-2y)^2$, use the formula for the square of a difference:
$$(a-b)^2 = a^2 - 2ab + b^2.$$
Here, $a = 5$ and $b = 2y$. Calculate each term:
$$a^2 = 5^2 = 25, \quad -2ab = -2(5)(2y) = -20y, \quad \text{and} \quad (2y)^2 = 4y^2.$$
Thus,
$$(5-2y)^2 = 25 - 20y + 4y^2.$$
(It is also acceptable to write this in standard form as $4y^2 - 20y + 25$.)

5. **Solution:** To expand $(6x + 8)(6x - 8)$, use the identity for the product of a sum and a difference:
$$(a + b)(a - b) = a^2 - b^2.$$
Here, let $a = 6x$ and $b = 8$. Compute:
$$a^2 = (6x)^2 = 36x^2 \quad \text{and} \quad b^2 = 8^2 = 64.$$
Therefore,
$$(6x + 8)(6x - 8) = 36x^2 - 64.$$

6. **Solution:** To multiply the numbers in scientific notation, multiply the coefficients and add the exponents:
$$(3.0 \times 10^5) \times (2.0 \times 10^3).$$
Multiply the coefficients:
$$3.0 \times 2.0 = 6.0.$$
Add the exponents:
$$10^5 \times 10^3 = 10^{5+3} = 10^8.$$
Therefore, the product in scientific notation is:
$$6.0 \times 10^8.$$

Chapter 7

Factoring: Greatest Common Factor and Simple Patterns

Extracting the Greatest Common Factor

In algebraic expressions composed of several terms, factoring begins with the identification of factors common to every term. The greatest common factor (GCF) is defined as the largest expression that divides each term exactly, without leaving a remainder. For numerical coefficients, the GCF is the greatest number that evenly divides all coefficients. For variables, the GCF includes each variable raised to the lowest exponent present in every term.

Consider the expression
$$12x^3y^2 + 18x^2y.$$

The coefficients 12 and 18 both have 6 as a common factor since $12 = 6 \times 2$ and $18 = 6 \times 3$. In the variable portion, the term $12x^3y^2$ contains x^3 while $18x^2y$ contains x^2; the common factor among them is x^2. Similarly, both terms contain at least one factor of y, with the smallest exponent being 1. Hence, the GCF is
$$6x^2y.$$

Extracting this factor from the original expression yields
$$12x^3y^2 + 18x^2y = 6x^2y\,(2xy + 3).$$

In another example, consider the expression
$$-8x^2 + 12x.$$
Both terms are divisible by 4 as a numerical factor and by x as a variable factor. Factoring out $-4x$ (which incorporates the negative sign so that the remaining expression has a positive leading coefficient) produces
$$-8x^2 + 12x = -4x(2x - 3).$$
This systematic approach, which involves identifying numerical commonalities and the lowest powers of variables, is essential for simplifying expressions and preparing them for further manipulations.

Recognizing Basic Factoring Patterns

Expressions that share a noticeable structure often permit further factoring after the extraction of the GCF. A common pattern arises in expressions where a monomial factor appears in each term. For example, consider the expression
$$ax + ay.$$
Here, the factor a is present in both terms and may be factored out to yield
$$ax + ay = a(x + y).$$
Another frequent scenario involves expressions where every term includes a numerical constant in common. For instance, the expression
$$7b + 14$$
can be factored by extracting the constant 7,
$$7b + 14 = 7(b + 2).$$

When variables with different exponents are involved, the common factor includes the variable raised to the smallest exponent found in the terms. This observation is crucial when simplifying polynomial expressions, as the process reduces complexity and often reveals further structural symmetries. In addition, when expressions contain a negative sign common to all terms, factoring out a negative common factor can result in an expression with a more standard form. Each of these techniques relies on recognizing the underlying pattern shared by the terms.

Worked Examples

A systematic procedure for factoring begins with the identification of all factors present in each term. After the factors have been established, the GCF is determined by selecting the highest factor that is common to every term. The original expression is then rewritten as the product of this GCF and a simplified expression.

Example 1: Examine the expression

$$24x^4 - 36x^3 + 12x^2.$$

The coefficients 24, 36, and 12 share a common factor of 12. The variable factors x^4, x^3, and x^2 share a common factor of x^2 (the smallest power among them). Thus, the GCF is

$$12x^2.$$

Factoring out $12x^2$ produces

$$24x^4 - 36x^3 + 12x^2 = 12x^2\left(2x^2 - 3x + 1\right).$$

Example 2: Consider the expression

$$15ab - 10a^2b^2 + 5ab^3.$$

Each term contains a numerical factor, and the numbers 15, 10, and 5 share a common factor of 5. In addition, every term includes the variables a and b; the smallest powers are a^1 and b^1. Therefore, the GCF is

$$5ab.$$

Extracting the GCF results in

$$15ab - 10a^2b^2 + 5ab^3 = 5ab\left(3 - 2ab + b^2\right).$$

These examples illustrate the methodical process of factoring by first identifying the GCF and then rewriting the expression as the product of this common factor and a reduced polynomial. Recognizing the basic patterns in each term leads to more efficient simplification and lays a foundation for further algebraic manipulation.

Multiple Choice Questions

1. Which of the following is the greatest common factor (GCF) of the expression
$$12x^3y^2 + 18x^2y?$$

    (a) $6x^2y$

    (b) $6x^3y^2$

    (c) $12x^2y$

    (d) $12xy$

2. When factoring the expression
$$-8x^2 + 12x,$$
which common factor should be extracted to yield a simplified form with a positive leading coefficient?

    (a) $-4x$

    (b) $4x$

    (c) $-2x$

    (d) $2x$

3. In determining the GCF for a variable that appears in multiple terms, which exponent should be used?

    (a) The highest exponent present among the terms.

    (b) The lowest exponent present among the terms.

    (c) The exponent from the first term only.

    (d) The sum of the exponents from all terms.

4. Which of the following correctly factors the expression
$$ax + ay?$$

    (a) $a(x+y)$

    (b) $x(a+y)$

    (c) $a + xy$

    (d) $(a+x)y$

5. Consider the expression

$$7b + 14.$$

What is its greatest common factor?

(a) b

(b) 14

(c) 7

(d) $7(b+2)$

6. Which of the following is NOT a step in the systematic process of factoring by extracting the GCF?

(a) Identifying the common numerical and variable factors in all terms.

(b) Determining the greatest common factor.

(c) Rewriting the original expression as the product of the GCF and a simplified polynomial.

(d) Omitting the GCF and directly applying other factoring methods.

7. After extracting the GCF from a polynomial, what is the typical subsequent step for further simplifying the expression?

(a) Expanding the factored form to verify the original expression.

(b) Checking for additional factoring opportunities within the simplified polynomial.

(c) Substituting numerical values for the variables.

(d) Recombining the factored terms into a single term.

Answers:

1. **A:** $6x^2y$
    The coefficients 12 and 18 have 6 as a common divisor. For the variables, x^3 and x^2 share x^2 (the smallest power) and y^2 and y share y. Hence, the GCF is $6x^2y$.

2. **A:** $-4x$
    Extracting $-4x$ from $-8x^2 + 12x$ gives $-4x(2x - 3)$, which is preferred because it makes the leading coefficient inside the parentheses positive.

3. **B: The lowest exponent present among the terms.**
   When factoring variables, you must take the variable raised to the smallest exponent found in every term to ensure it divides each term exactly.

4. **A:** $a(x+y)$
   Both terms in the expression $ax + ay$ contain the common factor a, which factors out to give $a(x+y)$.

5. **C: 7**
   In the expression $7b + 14$, the numbers 7 and 14 share 7 as their greatest common factor.

6. **D: Omitting the GCF and directly applying other factoring methods.**
   A systematic approach to factoring begins with identifying and extracting the GCF. Skipping this step is not a part of the standard factoring process.

7. **B: Checking for additional factoring opportunities within the simplified polynomial.**
   Once the GCF has been factored out, the remaining expression should be examined to see if it can be factored further, simplifying the expression even more.

Practice Problems

1. Factor the following expression:
$$30a^3b + 45a^2b^2 - 15ab$$

2. Factor the following expression:
$$-50y^3 + 20y^2$$

3. Factor the following expression:
$$6xy + 12xz$$

4. Factor the following expression:
$$14x^3 - 21x^2 + 7x$$

5. Factor the following expression:
$$4x^5 - 8x^3y + 12x^2y^2$$

6. Factor the following expression:
$$-16a^3b + 24a^2b^2 - 8ab^3$$

Answers

1. Factor the following expression:
$$30a^3b + 45a^2b^2 - 15ab.$$

   Solution:

   First, we identify the greatest common factor (GCF). For the numerical coefficients 30, 45, and 15, the GCF is 15. For the variable a, the smallest power among a^3, a^2, and a is a, and for b, the smallest power among b, b^2, and b is b. Thus, the overall GCF is
$$15ab.$$

Dividing each term by $15ab$:

$$\frac{30a^3b}{15ab} = 2a^2, \quad \frac{45a^2b^2}{15ab} = 3ab, \quad \frac{-15ab}{15ab} = -1.$$

Therefore, the expression factors as:

$$15ab\left(2a^2 + 3ab - 1\right).$$

2. Factor the following expression:

$$-50y^3 + 20y^2.$$

Solution:

The coefficients -50 and 20 have a common factor of 10. Factoring out a negative sign gives a positive leading term inside the parentheses, and both terms contain at least y^2. Thus, we factor out

$$-10y^2.$$

Dividing each term by $-10y^2$:

$$\frac{-50y^3}{-10y^2} = 5y, \quad \frac{20y^2}{-10y^2} = -2.$$

Thus, the expression factors as:

$$-10y^2\left(5y - 2\right).$$

3. Factor the following expression:

$$6xy + 12xz.$$

Solution:

The numerical coefficients 6 and 12 have a GCF of 6. Both terms contain the variable x, so the overall GCF is

$$6x.$$

Dividing each term by $6x$:

$$\frac{6xy}{6x} = y, \quad \frac{12xz}{6x} = 2z.$$

Hence, the factored form is:

$$6x\left(y + 2z\right).$$

4. Factor the following expression:
$$14x^3 - 21x^2 + 7x.$$

Solution:

For the coefficients 14, 21, and 7, the GCF is 7. Since each term contains at least one x, we factor out $7x$:
$$7x.$$

Dividing each term by $7x$:
$$\frac{14x^3}{7x} = 2x^2, \quad \frac{-21x^2}{7x} = -3x, \quad \frac{7x}{7x} = 1.$$

Therefore, the expression factors as:
$$7x\left(2x^2 - 3x + 1\right).$$

5. Factor the following expression:
$$4x^5 - 8x^3y + 12x^2y^2.$$

Solution:

The coefficients 4, 8, and 12 share a GCF of 4. For the variable x, the smallest power in the terms is x^2, and since y does not appear in every term, we do not factor it. Thus, the overall GCF is:
$$4x^2.$$

Dividing each term by $4x^2$:
$$\frac{4x^5}{4x^2} = x^3, \quad \frac{-8x^3y}{4x^2} = -2xy, \quad \frac{12x^2y^2}{4x^2} = 3y^2.$$

Thus, the factored expression is:
$$4x^2\left(x^3 - 2xy + 3y^2\right).$$

6. Factor the following expression:
$$-16a^3b + 24a^2b^2 - 8ab^3.$$

Solution:

The coefficients 16, 24, and 8 have a GCF of 8. To obtain

a positive leading term inside the parentheses, factor out a negative sign along with the common factors. Since every term contains at least one a and one b, and the smallest powers are a and b, the overall GCF is:

$$-8ab.$$

Dividing each term by $-8ab$:

$$\frac{-16a^3b}{-8ab} = 2a^2, \quad \frac{24a^2b^2}{-8ab} = -3ab, \quad \frac{-8ab^3}{-8ab} = b^2.$$

Therefore, the expression factors as:

$$-8ab\left(2a^2 - 3ab + b^2\right).$$

Chapter 8

Factoring: Difference of Squares and Perfect Square Trinomials

Difference of Squares

An expression that fits the format a² b², where both a² and b² represent perfect squares, is classified as a difference of squares. The structure of this expression permits a direct factorization into the product (a b)(a + b). Verification of this factorization follows from the standard expansion:

$$(a - b)(a + b) = a^2 + ab - ab - b^2 = a^2 - b^2.$$

A necessary condition for the application of this method is that both terms in the original expression must be perfect squares. Consider an expression such as 9x² 25. The term 9x² is recognized as (3x)² and 25 as 5². Consequently, the expression factors into

$$(3x - 5)(3x + 5).$$

In cases where the coefficients or variable parts are not immediately apparent as perfect squares, careful examination is required to rewrite each term as an exact square. For instance, an expression like 16y² 49 can be observed as the difference between (4y)² and 7². The factorization then results in

$$(4y - 7)(4y + 7).$$

This technique not only simplifies the expression but also eases the subsequent solution of equations or further algebraic manipulation.

Perfect Square Trinomials

A perfect square trinomial possesses the characteristic form $a^2 + 2ab + b^2$ or $a^2 - 2ab + b^2$. This structure indicates that the trinomial may be factored as $(a + b)^2$ or $(a - b)^2$, respectively. The identification of a perfect square trinomial begins with verifying that the first term is a perfect square, and that the last term is also a perfect square. The middle term, in its entirety, must equal twice the product of the square roots of the first and third terms.

For example, consider the quadratic expression $x^2 + 6x + 9$. The term x^2 is recognized as $(x)^2$, 9 is identified as $(3)^2$, and the middle term 6x equals 2 times x times 3. With these observations, the expression factors neatly into

$$(x + 3)^2.$$

Another illustrative instance is given by the expression $4x^2 - 12x + 9$. Here the term $4x^2$ can be rewritten as $(2x)^2$ and 9 as $(3)^2$. The linear term –12x satisfies the condition for perfect square trinomials since it equals –2 multiplied by (2x) and 3. Accordingly, the factorization is

$$(2x - 3)^2.$$

The systematic approach involves decomposing the trinomial into its component perfect squares and verifying that the middle term aligns exactly with the sum or difference of twice the product of the square roots of the outer terms. When these conditions are met, the factorization provides a useful tool for simplifying expressions and solving quadratic equations.

This chapter emphasizes the importance of recognizing the structural patterns that enable direct factorization. The method of decomposing a difference of squares or a perfect square trinomial into its constituent factors streamlines many algebraic procedures and contributes to a greater understanding of polynomial behavior.

Multiple Choice Questions

1. Which of the following expressions is an example of a difference of squares?

    (a) $9x^2 \ 25$
    (b) $9x^2 + 25$
    (c) $9x^2 \ 20$
    (d) $9x^2 + 20$

2. Factor the expression $16y^2 \ 49$.

    (a) $(4y \ 7)(4y + 7)$
    (b) $(4y \ 7)^2$
    (c) $(8y \ 7)(2y + 7)$
    (d) $(16y \ 7)(y + 7)$

3. What is a necessary condition for an expression to be factored as a difference of squares?

    (a) The expression must include a variable.
    (b) Both terms must be perfect squares.
    (c) The expression must have three terms.
    (d) The coefficients must be prime numbers.

4. Which of the following options represents a perfect square trinomial?

    (a) $x^2 + 6x + 9$
    (b) $x^2 + 7x + 16$
    (c) $x^2 \ 6x + 9$
    (d) Both (a) and (c)

5. In a perfect square trinomial, what must the middle term equal?

    (a) The square of the sum of the square roots of the first and third terms.
    (b) Twice the product of the square roots of the first and third terms.

(c) Half the product of the square roots of the first and third terms.

(d) The difference of the squares of the first and third terms.

6. Factor the perfect square trinomial $4x^2 - 12x + 9$.

    (a) $(2x - 3)^2$

    (b) $(2x + 3)^2$

    (c) $(4x - 9)^2$

    (d) $(2x - 3)(2x + 3)$

7. Which technique is most efficient for factoring $9x^2 - 25$?

    (a) Factoring by grouping

    (b) Using the quadratic formula

    (c) Recognizing the expression as a difference of squares

    (d) Completing the square

Answers:

1. **A: $9x^2 - 25$**
   Explanation: The expression $9x^2 - 25$ is a difference of squares because $9x^2$ can be written as $(3x)^2$ and 25 as 5^2. The difference of these two squares factors as $(3x - 5)(3x + 5)$.

2. **A: $(4y - 7)(4y + 7)$**
   Explanation: Recognizing $16y^2$ as $(4y)^2$ and 49 as 7^2, the expression $16y^2 - 49$ fits the pattern $a^2 - b^2$ and factors directly to $(4y - 7)(4y + 7)$.

3. **B: Both terms must be perfect squares**
   Explanation: For an expression to be factored as a difference of squares, each term must be a perfect square so that they can be rewritten in the form a^2 and b^2, allowing the factorization $(a - b)(a + b)$.

4. **D: Both (a) and (c)**
   Explanation: The expressions in (a) and (c) are perfect square trinomials. In (a), $x^2 + 6x + 9$ factors as $(x + 3)^2$, and in (c), $x^2 - 6x + 9$ factors as $(x - 3)^2$. Option (b) does not fit the perfect square pattern.

5. **B: Twice the product of the square roots of the first and third terms**
   Explanation: A perfect square trinomial must have a middle term equal to 2ab, where a and b are the square roots of the first and third terms, respectively. This condition ensures that the trinomial factors neatly as $(a \pm b)^2$.

6. **A: $(2x\ 3)^2$**
   Explanation: The expression $4x^2\ 12x + 9$ is a perfect square trinomial because $4x^2$ can be written as $(2x)^2$ and 9 as 3^2, while the middle term $12x$ matches $2(2x)(3)$. Therefore, it factors as $(2x\ 3)^2$.

7. **C: Recognizing the expression as a difference of squares**
   Explanation: The expression $9x^2\ 25$ is most efficiently factored by identifying it as a difference of squares since $9x^2 = (3x)^2$ and $25 = 5^2$, allowing immediate factorization to $(3x\ 5)(3x + 5)$ without more advanced techniques.

Practice Problems

1. Factor the expression:
$$9x^2 - 25$$

2. Factor the expression:
$$16y^2 - 49$$

3. Factor the perfect square trinomial:
$$x^2 + 6x + 9$$

4. Factor the perfect square trinomial:
$$4x^2 - 12x + 9$$

5. Factor the expression:
$$25 - 4a^2$$

6. Factor the perfect square trinomial:
$$9x^2 + 6x + 1$$

Answers

1. For the expression:
   $$9x^2 - 25$$
   Solution: Notice that $9x^2$ is a perfect square since $(3x)^2 = 9x^2$ and 25 is a perfect square because $5^2 = 25$. Hence, the expression fits the difference of squares form:
   $$a^2 - b^2 = (a - b)(a + b),$$
   where $a = 3x$ and $b = 5$. Therefore,
   $$9x^2 - 25 = (3x - 5)(3x + 5).$$

2. For the expression:
   $$16y^2 - 49$$
   Solution: Recognize that $16y^2 = (4y)^2$ and $49 = 7^2$. This allows us to write the expression as:
   $$(4y)^2 - 7^2.$$
   Applying the difference of squares formula:
   $$a^2 - b^2 = (a - b)(a + b),$$
   with $a = 4y$ and $b = 7$, we factor the expression as:
   $$16y^2 - 49 = (4y - 7)(4y + 7).$$

3. For the perfect square trinomial:
$$x^2 + 6x + 9$$
   Solution: Observe that x^2 is the square of x and 9 is the square of 3. The middle term $6x$ can be expressed as $2 \cdot x \cdot 3$. This matches the standard form:
$$a^2 + 2ab + b^2 = (a+b)^2,$$
   where $a = x$ and $b = 3$. Thus, we have:
$$x^2 + 6x + 9 = (x+3)^2.$$

4. For the perfect square trinomial:
$$4x^2 - 12x + 9$$
   Solution: Note that $4x^2 = (2x)^2$ and $9 = 3^2$. The middle term $-12x$ is equal to $-2 \cdot (2x) \cdot 3$, which fits the form:
$$a^2 - 2ab + b^2 = (a-b)^2,$$
   with $a = 2x$ and $b = 3$. Therefore,
$$4x^2 - 12x + 9 = (2x-3)^2.$$

5. For the expression:
$$25 - 4a^2$$
   Solution: Recognize that $25 = 5^2$ and $4a^2 = (2a)^2$. The expression is then of the form:
$$a^2 - b^2,$$
   with $a = 5$ and $b = 2a$. Using the difference of squares formula,
$$25 - 4a^2 = 5^2 - (2a)^2 = (5 - 2a)(5 + 2a).$$

6. For the perfect square trinomial:
$$9x^2 + 6x + 1$$
   Solution: Notice that $9x^2 = (3x)^2$ and $1 = 1^2$. The middle term $6x$ equals $2 \cdot 3x \cdot 1$, which fits the perfect square pattern:
$$a^2 + 2ab + b^2 = (a+b)^2,$$
   where $a = 3x$ and $b = 1$. Therefore,
$$9x^2 + 6x + 1 = (3x+1)^2.$$

Chapter 9

Factoring: Grouping and Quadratic Trinomials

Factoring by Grouping

The method of factoring by grouping is a powerful strategy for simplifying polynomials that contain four or more terms. In many cases, rearranging the terms of an expression into two or more groups reveals a common factor in each group. The process relies on the distributive property to factor out these common elements and then extract a common binomial factor.

Consider an expression with four terms in the form

$$ax + ay + bx + by.$$

Grouping the first two terms and the last two terms produces

$$(ax + ay) + (bx + by).$$

Factoring out the common factor a from the first group and b from the second group yields

$$a(x + y) + b(x + y).$$

Since $(x + y)$ appears in both terms, it can be factored out by applying the distributive property in reverse to obtain

$$(x + y)(a + b).$$

When the terms do not immediately present a common grouping, careful rearrangement may reveal hidden factors. For example, take the polynomial

$$x^3 + 2x^2 + 3x + 6.$$

By grouping the terms as $(x^3 + 2x^2)$ and $(3x + 6)$, a common factor in each group becomes evident:

$$x^2(x + 2) + 3(x + 2).$$

The binomial $(x+2)$ can now be factored from both groups, leading to

$$(x + 2)(x^2 + 3).$$

This example demonstrates how grouping terms appropriately can transform a seemingly complex polynomial into a product of simpler expressions.

Factoring Quadratic Trinomials

A quadratic trinomial is generally expressed in the standard form

$$ax^2 + bx + c,$$

where a, b, and c are constants. When the leading coefficient a is equal to 1, the process of factoring is simplified. The objective is to identify two numbers m and n such that

$$m + n = b \quad \text{and} \quad m \cdot n = c.$$

This allows the trinomial to be written as

$$(x + m)(x + n).$$

When the leading coefficient a is not equal to 1, a common strategy involves the product ac. Two numbers are sought that multiply to ac and add to b. Once these numbers are determined, the middle term bx of the quadratic is split into two terms, which enables the application of the grouping method.

For instance, consider the quadratic trinomial

$$2x^2 + 7x + 3.$$

The product ac is calculated as
$$(2)(3) = 6,$$
and the task is to find two numbers that multiply to 6 and add to 7. The numbers 6 and 1 satisfy these conditions because
$$6 \times 1 = 6 \quad \text{and} \quad 6 + 1 = 7.$$
Splitting the middle term $7x$ into $6x$ and x results in
$$2x^2 + 6x + x + 3.$$
This expression is then divided into two groups:
$$(2x^2 + 6x) + (x + 3).$$
In the first group, $2x$ can be factored out:
$$2x(x + 3).$$
In the second group, the factor 1 is common:
$$1(x + 3).$$
Thus, the expression becomes
$$2x(x + 3) + 1(x + 3),$$
and the common binomial $(x + 3)$ can be factored, yielding
$$(x + 3)(2x + 1).$$

In cases where the quadratic trinomial is a perfect square, the first and last terms are perfect squares and the middle term is exactly twice the product of the square roots of these terms. An example of such a scenario is
$$x^2 + 6x + 9,$$
where x^2 is $(x)^2$, 9 is $(3)^2$, and the middle term $6x$ equals $2 \cdot x \cdot 3$. This pattern corresponds to the square of a binomial and can be factored as
$$(x + 3)^2.$$

Both grouping and quadratic factoring require a systematic approach. For grouping, it is essential to identify and extract common factors from properly arranged subsets of terms. For quadratic trinomials, whether the leading coefficient is 1 or not, careful identification of the factors that satisfy the relationships between the coefficients is crucial. These methods are fundamental techniques that simplify expressions and solve equations efficiently.

Multiple Choice Questions

1. Which property is primarily used in the factoring by grouping method?

   (a) Commutative Property
   (b) Associative Property
   (c) Distributive Property
   (d) Identity Property

2. In the expression
$$ax + ay + bx + by,$$
   after grouping the terms and factoring common factors, what is the factored form?

   (a) $(a+b)(x+y)$
   (b) $(ax+bx)(ay+by)$
   (c) $a(x+b) + y(x+b)$
   (d) $(a+y)(b+x)$

3. When factoring a quadratic trinomial of the form
$$x^2 + bx + c,$$
   which conditions must the two numbers m and n satisfy?

   (a) $m+n = b$ and $m \cdot n = c$
   (b) $m+n = c$ and $m \cdot n = b$
   (c) $m+n = a$ and $m \cdot n = c$
   (d) $m+n = b$ and $m \cdot n = a$

4. For quadratic trinomials where the leading coefficient a is not 1, which product is used to split the middle term?

   (a) $a+c$
   (b) $a \times c$
   (c) $a+b$
   (d) $b \times c$

5. Given the quadratic trinomial
$$2x^2 + 7x + 3,$$
which pair of terms is used to split the middle term during the factoring process?

    (a) $3x$ and $4x$
    (b) $6x$ and x
    (c) $2x$ and $5x$
    (d) $7x$ and 0

6. Which of the following is a perfect square quadratic trinomial as discussed in the chapter?

    (a) $x^2 + 6x + 9$
    (b) $x^2 + 5x + 6$
    (c) $x^2 + 4x + 3$
    (d) $2x^2 + 6x + 9$

7. Why might it be necessary to rearrange the terms of a polynomial before applying the grouping method?

    (a) To change the sign of some terms
    (b) To create groups that share a common factor
    (c) To reduce the degree of the polynomial
    (d) To simplify the coefficients directly

Answers:

1. **C: Distributive Property**
   Explanation: In the factoring by grouping method, the distributive property is used in reverse to factor out a common binomial from grouped terms.

2. **A:** $(a + b)(x + y)$
   Explanation: Grouping the terms as $(ax + ay)$ and $(bx + by)$ allows you to factor out a from the first group and b from the second group, revealing the common binomial $(x + y)$. Thus, the expression factors as $(a + b)(x + y)$.

3. **A:** $m+n=b$ **and** $m \cdot n = c$
   Explanation: For a quadratic trinomial in the form x^2+bx+c, the goal is to find two numbers m and n whose sum is equal to b and whose product is equal to c.

4. **B:** $a \times c$
   Explanation: When the leading coefficient a is not 1, the product ac is used to determine the two numbers needed to split the middle term, aiding in the factoring process by grouping.

5. **B:** $6x$ **and** x
   Explanation: For the quadratic trinomial $2x^2 + 7x + 3$, the product ac is $2 \times 3 = 6$. The two numbers that multiply to 6 and add to 7 are 6 and 1. Thus, the middle term $7x$ is split into $6x$ and x.

6. **A:** $x^2 + 6x + 9$
   Explanation: A perfect square quadratic trinomial has its first and last terms as perfect squares and its middle term as twice the product of their square roots. The trinomial $x^2 + 6x + 9$ fits this description since it factors as $(x+3)^2$.

7. **B: To create groups that share a common factor**
   Explanation: Rearranging the terms of a polynomial can reveal groupings where a common factor exists. This strategy is essential for the grouping method to work efficiently when the common factors are not obvious in the original order.

Practice Problems

1. Factor the polynomial:
$$x^3 + 2x^2 + 3x + 6.$$

2. Factor the polynomial by grouping:
$$2x^3 - x^2 + 4x - 2.$$

3. Factor the quadratic trinomial:
$$x^2 + 5x + 6.$$

4. Factor the quadratic trinomial (with a leading coefficient not equal to 1):
$$6x^2 + 17x + 5.$$

5. Factor the perfect square quadratic:
$$4x^2 + 12x + 9.$$

6. Factor the polynomial completely:
$$2x^3 - 3x^2 - 8x + 12.$$

Answers

1. **Solution:**
   Begin by grouping the terms:
   $$(x^3 + 2x^2) + (3x + 6).$$
   Factor out the common factors in each group:
   $$x^2(x + 2) + 3(x + 2).$$
   Since $(x + 2)$ is common, factor it out:
   $$(x + 2)(x^2 + 3).$$

Therefore,
$$x^3 + 2x^2 + 3x + 6 = (x+2)(x^2+3).$$

2. **Solution:**
   Group the polynomial as follows:
   $$(2x^3 - x^2) + (4x - 2).$$
   Factor common factors from each group:
   $$x^2(2x-1) + 2(2x-1).$$
   Notice the common binomial $(2x-1)$. Factor it out:
   $$(2x-1)(x^2+2).$$
   Hence,
   $$2x^3 - x^2 + 4x - 2 = (2x-1)(x^2+2).$$

3. **Solution:**
   For the quadratic trinomial
   $$x^2 + 5x + 6,$$
   we look for two numbers that add to 5 and multiply to 6. The numbers 2 and 3 satisfy these conditions:
   $$2 + 3 = 5 \quad \text{and} \quad 2 \times 3 = 6.$$
   Therefore, it factors as:
   $$(x+2)(x+3).$$

4. **Solution:**
   Consider the quadratic
   $$6x^2 + 17x + 5.$$
   Multiply the leading coefficient and the constant term:
   $$6 \times 5 = 30.$$
   We need two numbers that add to 17 and multiply to 30. The numbers 15 and 2 work since:
   $$15 + 2 = 17 \quad \text{and} \quad 15 \times 2 = 30.$$

Rewrite the middle term:
$$6x^2 + 15x + 2x + 5.$$
Now, group the terms:
$$(6x^2 + 15x) + (2x + 5).$$
Factor out common factors from each group:
$$3x(2x + 5) + 1(2x + 5).$$
Factor the common binomial $(2x + 5)$:
$$(2x + 5)(3x + 1).$$
Thus,
$$6x^2 + 17x + 5 = (2x + 5)(3x + 1).$$

5. **Solution:**
The quadratic
$$4x^2 + 12x + 9$$
is a perfect square trinomial. Observe that:
$$4x^2 = (2x)^2 \quad \text{and} \quad 9 = 3^2,$$
and the middle term is:
$$12x = 2 \cdot (2x) \cdot 3.$$
Thus, it factors as:
$$(2x + 3)^2.$$

6. **Solution:**
Group the polynomial:
$$(2x^3 - 3x^2) + (-8x + 12).$$
Factor common factors from each group:
$$x^2(2x - 3) - 4(2x - 3).$$
Factor out the common binomial $(2x - 3)$:
$$(2x - 3)(x^2 - 4).$$
Notice that the quadratic $x^2 - 4$ is a difference of squares, which factors further:
$$x^2 - 4 = (x - 2)(x + 2).$$
Therefore, the complete factorization is:
$$2x^3 - 3x^2 - 8x + 12 = (2x - 3)(x - 2)(x + 2).$$

Chapter 10

Solving Linear Equations: Basic Techniques

Definition and Structure of One-Variable Linear Equations

A one-variable linear equation is one in which the variable appears to the first power and the equation represents a straight line when graphed on a coordinate plane. Such an equation is commonly expressed in the standard form

$$ax + b = c,$$

where a, b, and c are constant values with a nonzero. The structure of the equation emphasizes that the variable x is subject only to operations of multiplication by a constant and addition or subtraction by another constant. The goal is to find the unique value of x that produces equality between the left- and right-hand sides.

Properties of Equality and Inverse Operations

The systematic approach to solving linear equations is grounded in the fundamental properties of equality. These properties ensure that equivalent transformations on both sides of the equation do not disturb the balance of the equality. The following properties are central to the solution process:

Addition Property: If $A = B$, then $A + C = B + C$.

Subtraction Property: If $A = B$, then $A - C = B - C$.

Multiplication Property: If $A = B$, then
$$A \cdot C = B \cdot C \quad \text{(provided } C \neq 0\text{)}.$$

Division Property: If $A = B$, then $\dfrac{A}{C} = \dfrac{B}{C}$ (provided $C \neq 0$).

The technique of isolating the variable relies on these properties to remove unwanted constants or coefficients from the variable term. In effect, inverse operations are applied to "undo" the operations that have been performed on x.

Techniques for Isolating the Variable

The primary objective in solving a linear equation is to isolate the variable on one side of the equation. In an equation of the form

$$ax + b = c,$$

the operations that have been applied to x are undone by applying the inverse operations in the reverse order of operations. The process is typically executed in two major steps:

First, remove the constant term by applying the subtraction (or addition) property. Subtracting b from both sides of the equation yields

$$ax = c - b.$$

Next, eliminate the coefficient a through division. Dividing both sides by a produces

$$x = \frac{c - b}{a}.$$

This systematic reversal of operations ensures that the manipulation remains valid and the equality continues to hold at every stage of the solution process.

Step-by-Step Examples

In many cases, linear equations appear in an already simplified form; however, the method remains consistent regardless of the numerical values involved.

1 Example 1

Consider the equation
$$3x + 4 = 19.$$
The first step is to remove the constant term $+4$ by subtracting 4 from both sides:
$$3x = 19 - 4 = 15.$$
Next, divide both sides by the coefficient 3 to isolate x:
$$x = \frac{15}{3} = 5.$$

2 Example 2

Examine the equation
$$-2x + 6 = 0.$$
Begin by subtracting 6 from both sides to obtain
$$-2x = -6.$$
Then divide by -2 to solve for x:
$$x = \frac{-6}{-2} = 3.$$

3 Example 3

For the equation with variable terms on both sides,
$$2x + 5 = 3x - 2,$$
first remove the variable term from one side by subtracting $2x$ from both sides:
$$5 = x - 2.$$
Next, add 2 to both sides to isolate x:
$$x = 5 + 2 = 7.$$

Verification of Derived Solutions

After obtaining a solution for a linear equation, it is essential to validate the accuracy of the result by substituting the value back into the original equation. Consider the validation procedure in Example 1 where $x = 5$. Substituting in the equation

$$3x + 4 = 19,$$

leads to

$$3(5) + 4 = 15 + 4 = 19.$$

The equivalence of the left- and right-hand sides verifies the correctness of the solution. This substitution method applies universally to all linear equations and provides a reliable means of confirming that the applied techniques have produced a valid solution.

Multiple Choice Questions

1. Which of the following best describes the structure of a one-variable linear equation?

    (a) A quadratic equation where the variable is squared.

    (b) An equation in the form $ax + b = c$, where the variable appears only to the first power.

    (c) An equation written in slope-intercept form.

    (d) An equation involving fractions and radicals.

2. What is the first operation performed to solve an equation of the form $ax + b = c$?

    (a) Divide both sides by a.

    (b) Subtract b from both sides.

    (c) Add c to both sides.

    (d) Multiply both sides by b.

3. Which property of equality justifies dividing both sides of an equation by a nonzero number to isolate the variable?

    (a) Multiplication Property of Equality.

    (b) Division Property of Equality.

(c) Addition Property of Equality.

(d) Subtraction Property of Equality.

4. Consider the equation $-2x + 6 = 0$. What is the correct first step to solve this equation?

    (a) Add $2x$ to both sides.

    (b) Subtract 6 from both sides.

    (c) Divide both sides by -2.

    (d) Multiply both sides by -1.

5. Why is it essential to perform the same operation on both sides of an equation when solving it?

    (a) To preserve the balance of the equation.

    (b) To eliminate only the negative signs.

    (c) To remove the variable from one side entirely.

    (d) To change the structure of the equation.

6. For an equation in the form $ax + b = c$, what condition must be met before dividing both sides by a in order to isolate x?

    (a) a must be a positive number.

    (b) a must equal 1.

    (c) a must not equal 0.

    (d) a must be a fraction.

7. What is the primary purpose of substituting the solution back into the original equation after solving it?

    (a) To simplify the equation further.

    (b) To verify that the solution satisfies the original equation.

    (c) To convert the linear equation into a quadratic one.

    (d) To obtain additional possible solutions.

Answers:

1. **B: An equation in the form $ax + b = c$, where the variable appears only to the first power.**
   Explanation: A one-variable linear equation features the variable x raised only to the first power and is generally expressed as $ax + b = c$, highlighting its linear nature.

2. **B: Subtract b from both sides.**
   Explanation: The proper method to isolate the variable is to first eliminate the constant term b by subtracting it from both sides, which reverses the addition operation applied to x.

3. **B: Division Property of Equality.**
   Explanation: The Division Property of Equality allows us to divide both sides of an equation by the same nonzero constant (in this case, a) without altering the equality, thereby isolating x.

4. **B: Subtract 6 from both sides.**
   Explanation: For the equation $-2x + 6 = 0$, the first step is to subtract 6 from both sides to obtain $-2x = -6$, which then permits isolating x via division by -2.

5. **A: To preserve the balance of the equation.**
   Explanation: Performing an identical operation on both sides of an equation is crucial because it maintains the equality, ensuring that any transformation yields an equivalent equation with the same solution set.

6. **C: a must not equal 0.**
   Explanation: Dividing by zero is undefined; hence, the coefficient a must be nonzero to validly isolate x using the Division Property of Equality.

7. **B: To verify that the solution satisfies the original equation.**
   Explanation: Substituting the obtained value back into the original equation checks that all steps were executed correctly and confirms that the solution indeed makes the equation true.

Practice Problems

1. Solve the linear equation:
$$5x - 7 = 18$$

2. Solve the linear equation:
$$-3x + 9 = 0$$

3. Solve the linear equation:
$$2x + 5 = 3x - 2$$

4. Solve the linear equation:
$$\frac{1}{2}x + 4 = 10$$

5. Solve the linear equation:
$$-2x + 3 = 4x - 9$$

6. Solve the linear equation:
$$3(2x - 1) = 5x + 4$$

Answers

1. **Problem:** Solve
$$5x - 7 = 18.$$

   Solution:
   To isolate the variable, first add 7 to both sides of the equation:
   $$5x - 7 + 7 = 18 + 7,$$
   which simplifies to
   $$5x = 25.$$

Next, divide both sides by 5:
$$x = \frac{25}{5} = 5.$$

Explanation: We used the Addition Property of Equality to eliminate the constant term on the left, then applied the Division Property of Equality to isolate x. The final answer is $x = 5$.

2. **Problem:** Solve
$$-3x + 9 = 0.$$

Solution:
First, subtract 9 from both sides:
$$-3x + 9 - 9 = 0 - 9,$$
which gives
$$-3x = -9.$$
Then, divide both sides by -3:
$$x = \frac{-9}{-3} = 3.$$

Explanation: By subtracting 9, we removed the constant term from the left side. Dividing by the coefficient -3 isolated the variable. Thus, the solution is $x = 3$.

3. **Problem:** Solve
$$2x + 5 = 3x - 2.$$

Solution:
Start by eliminating the variable on one side. Subtract $2x$ from both sides:
$$2x + 5 - 2x = 3x - 2 - 2x,$$
resulting in
$$5 = x - 2.$$
Next, add 2 to both sides:
$$5 + 2 = x - 2 + 2,$$
which simplifies to
$$7 = x.$$

Explanation: We first gathered the x terms on one side by subtracting $2x$. Then, we removed the constant term on the right by adding 2. This procedure, following the inverse operations in reverse order, yields $x = 7$.

4. **Problem:** Solve
$$\frac{1}{2}x + 4 = 10.$$

Solution:
Begin by subtracting 4 from both sides to isolate the term containing x:
$$\frac{1}{2}x + 4 - 4 = 10 - 4,$$
which gives
$$\frac{1}{2}x = 6.$$
Multiply both sides by 2 to cancel the fraction:
$$x = 6 \cdot 2 = 12.$$

Explanation: The subtraction of 4 removed the constant term, and multiplying by 2 reversed the multiplication by $\frac{1}{2}$. Hence, the solution is $x = 12$.

5. **Problem:** Solve
$$-2x + 3 = 4x - 9.$$

Solution:
First, add $2x$ to both sides to gather the x-terms on one side:
$$-2x + 3 + 2x = 4x - 9 + 2x,$$
resulting in
$$3 = 6x - 9.$$
Next, add 9 to both sides:
$$3 + 9 = 6x - 9 + 9,$$
which simplifies to
$$12 = 6x.$$
Finally, divide both sides by 6:
$$x = \frac{12}{6} = 2.$$

Explanation: By transferring the x-terms and constants appropriately, we applied the Addition Property of Equality twice, then used the Division Property of Equality to isolate x. Therefore, the answer is $x = 2$.

6. **Problem:** Solve
$$3(2x - 1) = 5x + 4.$$

Solution:
First, distribute the 3 on the left side:
$$6x - 3 = 5x + 4.$$

Next, subtract $5x$ from both sides to collect like terms:
$$6x - 5x - 3 = 5x - 5x + 4,$$

which simplifies to
$$x - 3 = 4.$$

Then, add 3 to both sides:
$$x - 3 + 3 = 4 + 3,$$

resulting in
$$x = 7.$$

Explanation: We first applied the distributive property to eliminate the parentheses and then used inverse operations to move variable and constant terms to opposite sides. After combining like terms and isolating x, we obtained $x = 7$.

Chapter 11

Solving Linear Equations with Variables on Both Sides

Structure and Characteristics of Two-Sided Equations

Linear equations that include variables on both sides of the equality sign exhibit a structure in which terms containing the variable appear on each side. A general representation for these equations is
$$ax + b = cx + d,$$
where a, b, c, and d are constants and the variable x is raised only to the first power. The presence of the variable in two distinct locations necessitates a systematic approach that first consolidates the variable terms and then separates the constant terms. Each transformation applied to one side of the equation is matched by an identical operation on the other side in order to uphold the balance inherent in the equality.

Fundamental Properties and Principles of Equality

The solution of equations with variables on both sides is based on the foundational properties of equality. The Addition and Subtraction Properties allow movement of terms from one side of the equation to the other, while the Multiplication and Division Properties facilitate the removal of coefficients from variable terms. In particular, by subtracting or adding appropriate terms, like terms may be grouped together, and then division or multiplication isolates the variable. Each operation is performed under the guiding principle that the equality must remain balanced at every step.

Methodical Approach for Isolating the Variable

Solving equations that feature variables on both sides requires a clear and methodical sequence of steps. Consider an equation of the general form
$$ax + b = cx + d.$$
The initial step involves transferring all variable terms to one side of the equation. This is typically accomplished by subtracting cx from both sides, leading to
$$ax - cx + b = d.$$
Following this, the constant term on the side with the variable is removed by subtracting b from both sides, resulting in
$$(a - c)x = d - b.$$
Assuming that $a \neq c$ so that division by zero is avoided, dividing both sides by $(a - c)$ isolates the variable:
$$x = \frac{d - b}{a - c}.$$
This process utilizes inverse operations in the reverse order of the operations originally applied to the variable, ensuring that the structure of the original equation is methodically reversed until the value of x is fully isolated.

Worked Examples

1 Example 1

Examine the equation
$$3x + 2 = 5x - 6.$$
The solution begins by removing the variable term from one side through subtraction. Subtracting $3x$ from both sides results in
$$3x + 2 - 3x = 5x - 6 - 3x,$$
which simplifies to
$$2 = 2x - 6.$$
Next, adding 6 to both sides yields
$$2 + 6 = 2x - 6 + 6,$$
or equivalently,
$$8 = 2x.$$
Finally, dividing both sides by 2 provides the solution:
$$x = \frac{8}{2} = 4.$$

2 Example 2

Consider the equation
$$-4x + 7 = 2x - 5.$$
To gather the variable terms on one side, add $4x$ to both sides:
$$-4x + 7 + 4x = 2x - 5 + 4x,$$
which simplifies to
$$7 = 6x - 5.$$
Adding 5 to both sides gives
$$7 + 5 = 6x - 5 + 5,$$
thus,
$$12 = 6x.$$
Dividing both sides by 6 isolates x:
$$x = \frac{12}{6} = 2.$$

3 Example 3

Examine the equation
$$5x - 9 = 3x + 3.$$

The initial step is to consolidate like terms by subtracting $3x$ from both sides:
$$5x - 3x - 9 = 3x - 3x + 3,$$
which leads to
$$2x - 9 = 3.$$

Adding 9 to both sides shifts the constant term:
$$2x - 9 + 9 = 3 + 9,$$
yielding
$$2x = 12.$$

Finally, division by 2 isolates the variable:
$$x = \frac{12}{2} = 6.$$

Verification of Solutions

The verification of a solution is an essential step and involves substituting the obtained value back into the original equation to confirm that both sides remain equivalent. By replacing x with its computed value, each side of the equation is recalculated. If the resulting expressions on the left and right are equal, the solution is validated. This process not only confirms the correctness of the arithmetic but also reinforces the systematic nature of solving equations with variables on both sides.

Multiple Choice Questions

1. Which property of equality guarantees that performing the same operation on both sides of an equation preserves its balance?

    (a) Reflexive Property

    (b) Symmetric Property

(c) Addition (or Subtraction) Property of Equality

(d) Distributive Property

2. In solving a linear equation such as
$$ax + b = cx + d,$$
what is typically the first step to isolate the variable?

(a) Add b to both sides.

(b) Subtract cx from both sides.

(c) Divide both sides by a.

(d) Multiply both sides by x.

3. For the equation
$$ax + b = cx + d \quad (a \neq c),$$
what is the correct final expression for x after isolating the variable?

(a) $x = \frac{b-d}{a-c}$

(b) $x = \frac{d-b}{a-c}$

(c) $x = \frac{d+b}{a-c}$

(d) $x = \frac{d-b}{a+c}$

4. Which of the following is an essential step in verifying a solution to a two-sided linear equation?

(a) Substituting the solution back into the original equation.

(b) Comparing the coefficients of x on both sides.

(c) Reordering the terms to check commutativity.

(d) Graphing the equation and finding the intersection point.

5. When solving the equation
$$3x + 2 = 5x - 6,$$
which sequence of operations correctly isolates x?

(a) Subtract $3x$ from both sides, add 6 to both sides, then divide by 2.

(b) Subtract $5x$ from both sides, add 2 to both sides, then divide by 3.

(c) Add $3x$ to both sides, subtract 2 from both sides, then divide by 5.

(d) Subtract 2 from both sides, subtract $5x$ from both sides, then divide by 3.

6. What is the consequence of applying an operation to only one side of an equation while solving it?

   (a) The equation remains balanced.
   (b) The solution will still be correct.
   (c) The balance of the equation is lost, possibly leading to an incorrect solution.
   (d) The constant terms are eliminated automatically.

7. In the equation
$$-4x + 7 = 2x - 5,$$
which operation correctly groups the variable terms on one side?

   (a) Subtract 7 from both sides.
   (b) Add $4x$ to both sides.
   (c) Subtract $2x$ from both sides.
   (d) Add 5 to both sides.

Answers:

1. **C: Addition (or Subtraction) Property of Equality**
   This property ensures that when you add or subtract the same amount on both sides of an equation, the equality remains balanced.

2. **B: Subtract cx from both sides**
   The first step is to move all variable terms to one side by subtracting cx from both sides, which begins the process of isolating x.

3. **B:** $x = \frac{d-b}{a-c}$
   After moving the variable and constant terms appropriately (subtracting cx then b from both sides), dividing by $(a - c)$ (assuming $a \neq c$) yields this solution.

4. **A: Substituting the solution back into the original equation**
   Verifying the solution requires plugging the found value of x back into the original equation to ensure both sides are equal.

5. **A: Subtract $3x$ from both sides, add 6 to both sides, then divide by 2**
   This sequence first consolidates the x-terms on one side, then moves the constant term, and finally divides by the coefficient of x to isolate the variable.

6. **C: The balance of the equation is lost, possibly leading to an incorrect solution**
   Operations must be applied to both sides of the equation; failing to do so disrupts the balance and generally results in an invalid solution.

7. **B: Add $4x$ to both sides**
   Adding $4x$ to both sides moves all x-terms to one side, which is the correct initial step in isolating the variable in the given equation.

Practice Problems

1. Solve the following equation:

$$3x + 2 = 5x - 6$$

2. Solve the following equation:

$$-4x + 7 = 2x - 5$$

3. Solve the following equation:
$$5x - 9 = 3x + 3$$

4. Solve the following equation:
$$\frac{1}{2}x - 4 = \frac{1}{3}x + 2$$

5. Solve the following equation:
$$4(x - 2) = 2(2x + 1)$$

6. Solve the following equation:
$$-3(2x - 5) = 6 - 4x$$

Answers

1. For the equation:
$$3x + 2 = 5x - 6$$

   Solution:
   The first step is to eliminate the variable on one side. Subtract 3x from both sides:
   $$3x + 2 - 3x = 5x - 6 - 3x$$
   which simplifies to:
   $$2 = 2x - 6.$$
   Next, add 6 to both sides to isolate the term with x:
   $$2 + 6 = 2x - 6 + 6,$$
   giving:
   $$8 = 2x.$$
   Finally, divide both sides by 2:
   $$x = \frac{8}{2} = 4.$$
   Therefore, the solution is **x = 4**.

2. For the equation:
$$-4x + 7 = 2x - 5$$
Solution:
Begin by adding 4x to both sides to bring the variable terms together:
$$-4x + 7 + 4x = 2x - 5 + 4x,$$
which simplifies to:
$$7 = 6x - 5.$$
Next, add 5 to both sides:
$$7 + 5 = 6x - 5 + 5,$$
resulting in:
$$12 = 6x.$$
Divide both sides by 6 to isolate x:
$$x = \frac{12}{6} = 2.$$
Thus, the solution is **x = 2**.

3. For the equation:
$$5x - 9 = 3x + 3$$
Solution:
Subtract 3x from both sides:
$$5x - 9 - 3x = 3x + 3 - 3x,$$
which simplifies to:
$$2x - 9 = 3.$$
Next, add 9 to both sides:
$$2x - 9 + 9 = 3 + 9,$$
yielding:
$$2x = 12.$$
Finally, divide by 2:
$$x = \frac{12}{2} = 6.$$
Therefore, the solution is **x = 6**.

4. For the equation:
$$\frac{1}{2}x - 4 = \frac{1}{3}x + 2$$

Solution:
Start by subtracting $\frac{1}{3}x$ from both sides to collect like terms:
$$\frac{1}{2}x - \frac{1}{3}x - 4 = 2.$$

To combine $\frac{1}{2}x$ and $-\frac{1}{3}x$, find a common denominator:
$$\frac{3}{6}x - \frac{2}{6}x = \frac{1}{6}x.$$

Thus, the equation becomes:
$$\frac{1}{6}x - 4 = 2.$$

Add 4 to both sides:
$$\frac{1}{6}x = 6.$$

Finally, multiply both sides by 6 to solve for x:
$$x = 6 \times 6 = 36.$$

Hence, the solution is **x = 36**.

5. For the equation:
$$4(x - 2) = 2(2x + 1)$$

Solution:
Start by expanding both sides of the equation:
$$4(x - 2) = 4x - 8 \quad \text{and} \quad 2(2x + 1) = 4x + 2.$$

This gives us:
$$4x - 8 = 4x + 2.$$

Next, subtract 4x from both sides to eliminate the variable:
$$4x - 8 - 4x = 4x + 2 - 4x,$$

resulting in:
$$-8 = 2.$$

Since this is a contradiction, there is no value of x that satisfies the equation. Therefore, the equation has **no solution**.

6. For the equation:
$$-3(2x - 5) = 6 - 4x$$

Solution:
Begin by distributing -3 on the left side:
$$-3(2x) + (-3)(-5) = -6x + 15.$$

The equation becomes:
$$-6x + 15 = 6 - 4x.$$

Next, add 6x to both sides to gather the x-terms:
$$-6x + 15 + 6x = 6 - 4x + 6x,$$

which simplifies to:
$$15 = 6 + 2x.$$

Subtract 6 from both sides:
$$15 - 6 = 2x,$$

giving:
$$9 = 2x.$$

Finally, divide both sides by 2:
$$x = \frac{9}{2}.$$

Hence, the solution is $\mathbf{x = \frac{9}{2}}$.

Chapter 12

Applications of Linear Equations

Translating Real-Life Scenarios into Mathematical Models

In many everyday situations, relationships between quantities can be expressed through linear equations. The process begins with a careful examination of the scenario to identify the relationships that exist between unknown and known quantities. Quantitative elements, such as a fixed fee or a constant rate, are represented by numerical constants, while the unknown quantity is designated as a variable. When these components are combined in an arithmetic relationship, the result is a model that takes a form similar to

$$Ax + B,$$

where the coefficient A indicates the rate of change and B represents the fixed component. This method of translation provides a concise means to express complex real-world relationships in a format that is amenable to algebraic manipulation.

1 Identification of Variables and Constants

The first step in constructing a mathematical model involves discerning which quantities change and which remain constant. Quantities that do not vary within the context of the problem are assigned as constants, whereas the unknown quantity that is to be determined is designated as the variable. For instance, in a scenario involving a fixed service charge combined with an hourly rate, the total cost is influenced both by the fixed cost and by the product of the hourly rate and the number of hours. In this context, the fixed charge is a constant while the cost per hour multiplies the variable representing time, leading to an expression that combines both these elements.

2 Constructing the Algebraic Equation

After the relevant quantities have been identified, the next stage involves establishing the relationships between them. This is achieved by converting verbal descriptions into algebraic expressions. Phrases such as "the sum of," "increased by," or "multiplied by" correspond to addition, addition, and multiplication, respectively. In a typical application, if a total value is determined by a fixed amount and a contribution that varies with a particular measure, the model can be constructed as

$$\text{Total} = (\text{Rate}) \cdot (\text{Variable}) + (\text{Fixed Amount}).$$

An accurately formulated equation encapsulates all necessary elements of the scenario, allowing for subsequent algebraic manipulation.

Algebraic Techniques in Solving Word Problems

Once an equation has been established from a real-world scenario, standard algebraic techniques are employed to solve it. These techniques are founded on the properties of equality and include the systematic application of inverse operations to isolate the variable. The linear equation is manipulated through operations that maintain balance on both sides of the equality sign, a process that leads to the determination of the unknown quantity.

1 Application of Inverse Operations

Solving a linear equation typically requires the consolidation of like terms and the use of inverse operations. For example, when an equation is of the form

$$Ax + B = Cx + D,$$

the first step is to gather the variable terms on one side by subtracting one of the variable terms from both sides. After the variable terms have been consolidated and the constant terms have been rearranged, division by the coefficient of the variable isolates the unknown. Each step of this process is governed by the principle that performing the same operation on both sides of an equation preserves equality.

2 Verification and Interpretation of Solutions

Following the isolation and determination of the unknown variable, a verification process is essential. This is achieved by substituting the computed value back into the original equation. The equality must hold true after substitution, thereby confirming the accuracy of the model and the solution. The arithmetic validation reinforces the connection between the real-world scenario and its mathematical representation, ensuring that the derived solution accurately reflects the conditions that were originally described.

Illustrative Examples of Practical Applications

Linear equations find extensive application in various real-life contexts. The creation and solution of these equations provide insight into everyday problems such as financial planning, rate determination, and mixture calculations. The following examples highlight how linear models are embedded within practical scenarios.

1 Applications in Financial Calculations

Consider a situation where a service provider charges a fixed fee for an appointment along with an additional fee per hour of service. If the fixed fee is denoted by B and the hourly rate by A, and if x represents the number of hours of service, the total cost C is given by
$$C = Ax + B.$$

This model allows for the calculation of the total expense based on the duration of service and provides a straightforward method to compare options or estimate costs under varying conditions.

2 Applications in Rate Problems

Another common scenario involves problems related to distance, speed, and time. When an object moves at a constant speed, the distance d traveled is related to the time x by the equation

$$d = rx,$$

where r denotes the constant speed. By rearranging the equation, it is possible to solve for the time required to traverse a known distance or to determine the speed necessary to cover a distance within a specified time frame. This approach simplifies the analysis of motion and enhances the understanding of proportional relations.

3 Applications in Mixture Problems

Mixture problems often involve combining substances in different proportions to achieve a desired concentration or quantity. In such cases, linear equations are used to model the overall relationships between the components. If one substance is added at a rate proportional to a variable and the total required mixture is fixed, the resulting equation provides a clear representation of the balance between the contributors. The model enables the calculation of necessary amounts to sustain the desired proportion, thereby offering an effective tool for planning and resource allocation.

4 Integration of Analytical Methods

The versatility of linear equations lies in their ability to integrate analytical techniques with practical applications. From identifying the relevant parameters in a scenario to executing systematic algebraic manipulations, each step is critical for transforming a qualitative description into a quantitative analysis. By accurately modeling real-world problems with linear equations, the solutions obtained are directly applicable and provide meaningful insights into everyday situations.

Multiple Choice Questions

1. Which of the following best describes the role of the variable in a linear equation derived from a real-life scenario?

    (a) It represents a fixed quantity that does not change.

(b) It represents an unknown quantity that can vary.

(c) It represents the rate of change in the situation.

(d) It represents the total sum of the given components.

2. In a financial scenario where a service provider charges a fixed fee plus an hourly rate, which equation best represents the total cost, C?

   (a) C = (Fixed Fee) × (Hourly Rate) × (Number of Hours)
   (b) C = (Fixed Fee) + (Hourly Rate) × (Number of Hours)
   (c) C = (Hourly Rate) - (Fixed Fee) × (Number of Hours)
   (d) C = (Fixed Fee) ÷ [(Hourly Rate) × (Number of Hours)]

3. In the linear model Ax + B, what does the coefficient A typically represent?

   (a) The starting value or initial amount.
   (b) The rate at which the total quantity changes with respect to x.
   (c) The overall total of the measured quantity.
   (d) The fixed component in the modeled situation.

4. When verifying the solution of a linear equation obtained from a word problem, which step is essential?

   (a) Rearranging the equation into a different order.
   (b) Substituting the solution back into the original equation.
   (c) Graphing the equation on a Cartesian plane.
   (d) Changing the variables to see if the structure remains constant.

5. What is the primary purpose of applying inverse operations when solving linear equations from real-world scenarios?

   (a) To rewrite the equation into a more complex form.
   (b) To isolate the variable and compute its value.
   (c) To combine multiple equations into one comprehensive model.
   (d) To eliminate the constants from the equation.

6. If an object moves at a constant speed, which of the following equations correctly relates distance (d) to time (x) and speed (r)?

    (a) $d = r + x$
    (b) $d = r - x$
    (c) $d = r \times x$
    (d) $d = r \div x$

7. In mixture problems modeled by linear equations, the objective is primarily to:

    (a) Maximize the unknown quantity without restrictions.
    (b) Determine the correct proportions of substances to achieve a desired mixture.
    (c) Convert the problem into a financial scenario.
    (d) Estimate the square of the unknown variable.

Answers:

1. **B: It represents an unknown quantity that can vary.**
   Explanation: In word problems, the variable is used to denote an unknown value that needs to be determined, while constants account for fixed amounts.

2. **B: C = (Fixed Fee) + (Hourly Rate) × (Number of Hours)**
   Explanation: In this financial model, the total cost equals the fixed fee plus the product of the hourly rate and the number of hours, accurately reflecting the linear relationship.

3. **B: The rate at which the total quantity changes with respect to x.**
   Explanation: Within the model Ax + B, the coefficient A multiplies the variable x, indicating the rate of change or the per-unit contribution of the variable to the total.

4. **B: Substituting the solution back into the original equation.**
   Explanation: Verification is achieved by plugging the obtained solution into the original equation to ensure that both sides are equal, confirming the solution's correctness.

5. **B: To isolate the variable and compute its value.**
   Explanation: Inverse operations are applied to "undo" operations in the equation—such as division undoing multiplication—thus isolating the variable so that its value can be determined.

6. **C:** $d = r \times x$

   Explanation: This equation represents the basic relationship in rate problems, where distance is calculated as the product of the constant speed (rate) r and time x.

7. **B: Determine the correct proportions of substances to achieve a desired mixture.**
   Explanation: In mixture problems, linear equations are used to balance different components, ensuring that the proportion of each substance meets the requirements of the desired final mixture.

Practice Problems

1. A taxi service charges a fixed fare plus a per-mile rate. If the fixed fare is 3.50 and the charge per mile is 2.00, write a linear equation that represents the total cost C for traveling x miles. Also, calculate the cost for a journey of 10 miles.

2. A cell phone plan consists of a fixed monthly fee plus an additional fee per gigabyte of data used. With a monthly fee of 20 and a per-gigabyte fee of 5, write a linear equation that expresses the total monthly cost T in terms of the data usage x (in gigabytes). Then, find the total cost when 8 gigabytes are used.

3. Solve the following linear equation and verify your solution by substitution:
$$4x + 7 = 3x + 19$$

4. A car rental service charges a fixed fee plus a daily rate. If the fixed fee is 50 and the daily rate is 30, write an expression for the total rental cost C for renting a car for x days. Calculate the cost for a 5-day rental period.

5. A customer's cell phone bill includes a basic charge plus an extra charge of 2 per additional minute. If the customer used 20 extra minutes and the total bill was 65, set up a linear equation to find the basic charge B, and solve for B.

6. In a mixture problem, 50 liters of a solution with a 10 percent concentration is mixed with x liters of a 20 percent solution to obtain a mixture with a 15 percent concentration. Write a linear equation representing this situation and solve for x.

Answers

1. **Solution:** The total cost C consists of a variable cost that depends on the number of miles traveled plus a fixed fare. The linear equation is:
$$C = 2x + 3.50$$
For a journey of 10 miles, substitute x = 10:
$$C = 2(10) + 3.50 = 20 + 3.50 = 23.50$$
Therefore, the cost for 10 miles is 23.50.

2. **Solution:** The total monthly cost T is the sum of the fixed fee and the cost based on the data usage. The equation is:
$$T = 5x + 20$$
For a usage of 8 gigabytes, substitute x = 8:
$$T = 5(8) + 20 = 40 + 20 = 60$$
Thus, the total monthly cost when using 8 gigabytes is 60.

3. **Solution:** Begin with the equation:
$$4x + 7 = 3x + 19$$
Subtract 3x from both sides to collect the variable terms:
$$x + 7 = 19$$

Then subtract 7 from both sides:
$$x = 12$$

To verify, substitute x = 12 back into the original equation:
$$4(12) + 7 = 48 + 7 = 55$$

and
$$3(12) + 19 = 36 + 19 = 55$$

Since both sides equal 55, the solution x = 12 is correct.

4. **Solution:** The total rental cost C is made up of a fixed fee and a cost that depends on the number of rental days. The expression is:
$$C = 50 + 30x$$

For a 5-day rental, substitute x = 5:
$$C = 50 + 30(5) = 50 + 150 = 200$$

Therefore, the cost for renting the car for 5 days is 200.

5. **Solution:** Let B represent the basic charge. The extra charge for 20 minutes is:
$$2(20) = 40$$

The total bill is given by:
$$B + 40 = 65$$

To solve for B, subtract 40 from both sides:
$$B = 65 - 40 = 25$$

Hence, the basic charge is 25.

6. **Solution:** First, determine the amount of pure substance in the 50-liter solution with a 10 percent concentration:
$$0.10 \times 50 = 5$$

The x liters of the 20 percent solution contribute:
$$0.20x$$

The total volume of the mixture is:

$$50 + x$$

Since the final mixture has a 15 percent concentration, the total amount of pure substance in the mixture is:

$$0.15(50 + x)$$

Set up the equation:

$$5 + 0.20x = 0.15(50 + x)$$

Expand the right-hand side:

$$5 + 0.20x = 7.5 + 0.15x$$

Subtract 0.15x from both sides:

$$5 + 0.05x = 7.5$$

Subtract 5 from both sides:

$$0.05x = 2.5$$

Divide both sides by 0.05:

$$x = \frac{2.5}{0.05} = 50$$

Therefore, 50 liters of the 20 percent solution must be added.

Chapter 13

Solving Linear Inequalities

Definition and Fundamental Concepts

1 Inequality Symbols and Notation

Linear inequalities express a relationship in which one linear expression is either less than, greater than, or equal to another linear expression in an inclusive manner. The symbols $<$, $>$, $\leq$, and $\geq$ are used to denote these relationships. In these expressions, the boundary value may or may not be included in the solution set. A linear inequality shares many similarities with a linear equation; however, it establishes a range of values, rather than a single value, that satisfies the stated condition.

2 Relationship Between Linear Equations and Inequalities

The process of solving a linear inequality is analogous to that of solving a linear equation. The same arithmetic operations, including addition, subtraction, multiplication, and division, are applied to both sides of the inequality to isolate the variable. A difference arises when multiplying or dividing by a negative number, which necessitates a reversal of the inequality sign. This unique aspect distinguishes the manipulation of inequalities from that of equations, ensuring that the logical relationship between the expressions remains valid.

Techniques for Solving Linear Inequalities

1 Isolating the Variable

The initial approach to solving a linear inequality involves isolating the variable on one side of the inequality. This is accomplished by eliminating constant terms and coefficients that accompany the variable. For an inequality of the form $Ax + B < C$, subtracting B from both sides yields $Ax < C - B$. Once the variable term is isolated, further operations can be undertaken to solve for the variable itself.

2 Application of Inverse Operations

Inverse operations play a critical role in the process of solving inequalities. By applying addition or subtraction to remove constant terms and multiplication or division to eliminate coefficients, the variable is brought into isolation. Each operation is carefully performed on both sides of the inequality in order to preserve the original relationship between the expressions. This systematic manipulation allows for the gradual simplification of the inequality until a direct expression for the variable is obtained.

3 Handling Negative Coefficients

A notable consideration occurs when operations involve multiplication or division by a negative coefficient. In such cases, the direction of the inequality must be reversed to maintain the truth of the statement. For example, consider an inequality that contains $-Mx$; dividing both sides of the inequality by $-M$ results in the reversal of the inequality sign. This step is essential to correctly reflecting the effect that negative factors have on the magnitude of the variable's contribution.

Graphical Representation on the Number Line

1 Concept of the Number Line

The number line serves as an effective tool for visually representing the solution set of a linear inequality. It is a horizontal line marked with evenly spaced points corresponding to real numbers. This visual representation provides insight into the spatial relationships between values and the continuous nature of the solution set. The number line enables a clear demonstration of where the solutions lie relative to a specific boundary point.

2 Endpoints: Open and Closed Circles

When depicting the solution set on the number line, endpoints are distinguished by the use of open or closed circles. An open circle indicates that the endpoint is not part of the solution set, which is appropriate for inequalities involving the symbols $<$ or $>$. In contrast, a closed circle signifies that the endpoint is included in the solution, as seen in inequalities with the symbols $\leq$ or $\geq$. This notation effectively communicates whether the boundary value itself satisfies the inequality.

3 Shading the Interval

After determining the appropriate endpoint notation, the solution interval is portrayed by shading the region on the number line that corresponds to all values satisfying the inequality. For an inequality such as $x > a$, an open circle is placed at a and the region to the right is shaded to represent all values greater than a. Similarly, for an inequality like $x \leq b$, a closed circle is drawn at b with shading extending to the left, illustrating that every value less than or equal to b is part of the solution set.

Worked Examples with Step-by-Step Explanations

Consider the inequality:

$$3x - 5 < 7$$

The first step involves eliminating the constant term on the side with the variable. By adding 5 to both sides, the inequality becomes:
$$3x < 12$$
Subsequently, the variable is isolated by dividing both sides by 3. Since the division involves a positive number, the direction of the inequality remains unchanged:
$$x < 4$$
On the number line, an open circle is placed at 4, and the region extending to the left is shaded, indicating that all values less than 4 satisfy the inequality.

Next, consider an inequality that involves a negative coefficient:
$$-2x + 3 \geq 7$$
The initial step is to subtract 3 from both sides to remove the constant term:
$$-2x \geq 4$$
After isolating the variable term, division is performed on both sides by -2. Due to the negative divisor, the inequality sign is reversed, yielding:
$$x \leq -2$$
Graphically, this solution is represented on the number line by a closed circle at -2 and shading extending to the left, illustrating that all real numbers less than or equal to -2 constitute the solution.

The detailed approach in each example emphasizes the importance of methodical operations and adherence to arithmetic principles. Throughout the process, each step is carefully justified by the underlying properties of inequalities, ensuring that the final solution accurately reflects all conditions set forth by the original inequality.

Multiple Choice Questions

1. Which of the following best distinguishes a linear inequality from a linear equation?

    (a) A linear inequality always has exactly two solutions.

(b) A linear inequality represents a range of values rather than a single value.

   (c) A linear inequality uses the equals sign repeatedly.

   (d) A linear inequality involves quadratic expressions.

2. When solving a linear inequality, why must you reverse the inequality sign when multiplying or dividing both sides by a negative number?

   (a) Because negative numbers cancel out.

   (b) Because the order of the numbers on the number line reverses.

   (c) Because it simplifies the expression.

   (d) Because the inequality would otherwise have no solution.

3. In the inequality $-2x + 3 \geq 7$, what is the correct first step to isolate the variable?

   (a) Divide both sides by -2.

   (b) Add $2x$ to both sides.

   (c) Subtract 3 from both sides.

   (d) Multiply both sides by 2.

4. When graphing the solution to the inequality $x < 4$ on a number line, how is the endpoint at 4 represented?

   (a) With a closed circle, because 4 is included.

   (b) With an open circle, because 4 is not included.

   (c) With a bold dot, indicating the exact value.

   (d) With shading at 4 to indicate its exclusion.

5. What does the shading on a number line typically represent when graphing the solution to a linear inequality?

   (a) The values that do not satisfy the inequality.

   (b) The midpoint of the solution set.

   (c) The set of all values that satisfy the inequality.

   (d) Only the endpoint of the inequality.

6. For the inequality $3x - 5 < 7$, what is the final solution after correctly solving it?

(a) $x < 3$

(b) $x < 4$

(c) $x > 3$

(d) $x > 4$

7. Which of the following best describes the general process for solving linear inequalities?

    (a) Isolate the variable and perform the same operations as in solving a linear equation, with special consideration for multiplying or dividing by negative numbers.

    (b) Reverse the inequality sign at every step, regardless of the operation.

    (c) Only add or subtract constants without isolating the variable.

    (d) Multiply both sides by a constant to simplify the inequality.

Answers:

1. **B: A linear inequality represents a range of values rather than a single value.**
Explanation: While a linear equation yields one unique solution, a linear inequality defines a set or interval of values that satisfy the condition.

2. **B: Because the order of the numbers on the number line reverses.**
Explanation: Multiplying or dividing both sides of an inequality by a negative number reverses the direction of the inequality since the relative order of the numbers is inverted.

3. **C: Subtract 3 from both sides.**
Explanation: Removing the constant term that is added to the variable term is the first step in isolating the variable; after subtracting 3, the variable term is easier to manipulate.

4. **B: With an open circle, because 4 is not included.**
Explanation: In the inequality $x < 4$, the number 4 is not a solution; an open circle on the number line signifies that the endpoint is excluded from the solution set.

5. **C: The set of all values that satisfy the inequality.**
   Explanation: Shading on a number line indicates every number that meets the inequality criteria, thus visually representing the entire solution set.

6. **B:** $x < 4$
   Explanation: Adding 5 to both sides of the inequality $3x - 5 < 7$ results in $3x < 12$; dividing by 3 then gives $x < 4$.

7. **A: Isolate the variable and perform the same operations as in solving a linear equation, with special consideration for multiplying or dividing by negative numbers.**
   Explanation: The process for solving linear inequalities mirrors that for linear equations, but extra care must be taken to reverse the inequality sign when dealing with negative multipliers or divisors.

Practice Problems

1. Solve the following inequality:

$$4x - 7 \leq 9$$

2. Solve the following inequality:

$$-3x + 5 > 2$$

3. Solve the following inequality:
$$2(x-1) < 3x + 4$$

4. Solve the compound inequality:
$$-2 \leq 3x - 1 < 8$$

5. Solve the following inequality:
$$\frac{-x}{2} \geq 3$$

6. Explain why the inequality sign reverses when multiplying or dividing both sides of an inequality by a negative number.

Answers

1. **Problem:** Solve the inequality:
$$4x - 7 \leq 9$$

   Solution: Start by isolating the term with x. Add 7 to both sides:
   $$4x - 7 + 7 \leq 9 + 7 \implies 4x \leq 16$$

   Then divide both sides by 4 (since 4 is positive, the inequality sign remains unchanged):
   $$x \leq \frac{16}{4} \implies x \leq 4$$

   Explanation: We eliminated the constant by adding 7 to both sides and then divided by the positive number 4. Since dividing by a positive number does not change the direction of the inequality, the solution is all real numbers x such that x is less than or equal to 4.

2. **Problem:** Solve the inequality:
$$-3x + 5 > 2$$

   Solution: Begin by subtracting 5 from both sides:
   $$-3x + 5 - 5 > 2 - 5 \implies -3x > -3$$

   Next, divide both sides by -3. Remember that dividing by a negative number reverses the inequality sign:
   $$x < \frac{-3}{-3} \implies x < 1$$

Explanation: After isolating the term in x, we divided by -3. Since division by a negative value requires flipping the inequality sign, the final solution is $x < 1$.

3. **Problem:** Solve the inequality:
$$2(x - 1) < 3x + 4$$

Solution: First, distribute the 2:
$$2x - 2 < 3x + 4$$

Next, subtract $2x$ from both sides to collect like terms:
$$-2 < x + 4$$

Then subtract 4 from both sides:
$$-2 - 4 < x \implies -6 < x$$

This inequality can be rewritten as:
$$x > -6$$

Explanation: We distributed and then combined like terms to isolate x. The final inequality reveals that all numbers greater than -6 are solutions.

4. **Problem:** Solve the compound inequality:
$$-2 \leq 3x - 1 < 8$$

Solution: Treat the compound inequality as two separate inequalities.

First, solve:
$$-2 \leq 3x - 1$$

Add 1 to both sides:
$$-2 + 1 \leq 3x \implies -1 \leq 3x$$

Divide by 3:
$$x \geq -\frac{1}{3}$$

Next, solve:
$$3x - 1 < 8$$

Add 1 to both sides:
$$3x < 9$$
Divide by 3:
$$x < 3$$
Combine the two results:
$$-\frac{1}{3} \leq x < 3$$

Explanation: We split the compound inequality into two parts and solved each separately. The solution is the intersection of the two ranges: x must be greater than or equal to $-\frac{1}{3}$ and less than 3.

5. **Problem:** Solve the inequality:
$$\frac{-x}{2} \geq 3$$

Solution: Multiply both sides by 2 to eliminate the denominator:
$$-x \geq 6$$
Next, multiply both sides by -1 (remembering to reverse the inequality sign when multiplying by a negative number):
$$x \leq -6$$

Explanation: By first removing the fraction (multiplying by 2) and then multiplying by -1, we reversed the inequality sign. Thus, the solution is all x such that x is less than or equal to -6.

6. **Problem:** Explain why the inequality sign reverses when multiplying or dividing both sides of an inequality by a negative number.

Answer: Multiplying or dividing by a negative number reverses the direction of the inequality because it flips the order of the numbers on the number line. In essence, multiplying by a negative number reflects each number about zero. For example, if we have two numbers a and b such that $a < b$, then multiplying both by -1 gives $-a$ and $-b$. Since $-a$ is now greater than $-b$ (i.e., $-a > -b$), the inequality sign must be reversed to maintain the correct relational order. This reversal preserves the true comparison between the values after the sign change.

Chapter 14

Compound Inequalities and Absolute Value Equations

Compound Inequalities

1 Definition and Notation of Compound Inequalities

A compound inequality is formed by connecting two distinct inequality statements with either the conjunction "and" or the disjunction "or." When the connector used is "and," the inequality expresses that the variable must satisfy both conditions simultaneously. For example, an inequality written in the form

$$L \leq Ax + B < U$$

indicates that the linear expression $Ax + B$ is bounded below by L and strictly less than U. In contrast, an inequality connected by "or" represents a situation in which the variable can satisfy either one of the conditions. Special attention is given to the placement of the inequality symbols, as the symbols $\leq$ and $\geq$ denote that a boundary value is included in the solution set, whereas the symbols $<$ and $>$ denote that it is excluded.

2 Techniques for Solving Compound Inequalities

The resolution of a compound inequality relies on performing the same arithmetic operations on every segment of the inequality concurrently. Consider a standard compound inequality of the form

$$L \leq Ax + B < U.$$

The procedure often begins with the elimination of any constant term by adding or subtracting the same number from each part of the inequality. For instance, subtracting B from all three segments yields

$$L - B \leq Ax < U - B.$$

Subsequent operations involve isolating the term containing the variable. When the coefficient A is positive, division by A results in

$$\frac{L - B}{A} \leq x < \frac{U - B}{A}.$$

If the coefficient is negative, division requires the reversal of the inequality signs to maintain the proper relational order. The systematic application of these steps ensures that the solution correctly represents the set of values that simultaneously satisfy both portions of the compound inequality.

3 Worked Examples on Compound Inequalities

Consider the compound inequality

$$-2 \leq 3x - 1 < 8.$$

The initial step involves isolating the term containing the variable by adding 1 to each segment. This adjustment transforms the inequality into

$$-2 + 1 \leq 3x - 1 + 1 < 8 + 1,$$

leading to

$$-1 \leq 3x < 9.$$

To isolate x, division by the positive coefficient 3 is performed throughout the inequality. This division gives

$$-\frac{1}{3} \leq x < 3.$$

The resulting expression precisely delineates the set of values x that are greater than or equal to $-\frac{1}{3}$ and strictly less than 3.

Absolute Value Equations

1 Definition and Properties of Absolute Value

The absolute value of a number is defined as its distance from zero on the number line, regardless of direction. This property is denoted by two vertical bars, as in $|x|$. The key characteristic of the absolute value function is that it always returns a nonnegative quantity. For any real number x, the definition can be formally stated as

$$|x| = \begin{cases} x, & \text{if } x \geq 0, \\ -x, & \text{if } x < 0. \end{cases}$$

This piecewise definition is central to developing strategies for solving equations that include absolute value expressions, since the expression inside the absolute value symbol could correspond to both a positive and a negative scenario that produce the same nonnegative outcome.

2 Solving Equations Involving Absolute Values

Equations that involve absolute values require careful handling due to the dual nature of the absolute value function. An equation of the form

$$|Ax + B| = C,$$

where C is a nonnegative constant, is solved by considering the two distinct cases that satisfy the equation. The first case sets the expression inside the absolute value equal to C, yielding

$$Ax + B = C,$$

and the second case sets the expression equal to the opposite of C, yielding

$$Ax + B = -C.$$

Both resulting linear equations are then solved separately. It is important to note that if C is negative, the equation has no solution, since the absolute value cannot be negative. This two-case strategy facilitates the analysis of the equation and ensures that all potential solutions are determined.

3 Worked Examples on Absolute Value Equations

Examine the equation
$$|2x - 3| = 5.$$
The approach begins by considering the scenario where the expression inside the absolute value equals 5:
$$2x - 3 = 5.$$
Adding 3 to both sides results in
$$2x = 8,$$
and dividing by 2 yields
$$x = 4.$$
Next, the case in which the expression equals -5 is examined:
$$2x - 3 = -5.$$
Adding 3 to both sides produces
$$2x = -2,$$
and dividing by 2 results in
$$x = -1.$$
Therefore, the solutions to the equation $|2x - 3| = 5$ are $x = 4$ and $x = -1$, both of which satisfy the original equation.

In a scenario with more complex equations, the same principles apply. The absolute value expression is isolated on one side, and the equation is rewritten by considering both the positive and negative cases. This method provides a systematic framework for dealing with the inherent duality embedded in absolute value expressions.

Multiple Choice Questions

1. Which of the following best describes a compound inequality?

    (a) A single inequality using one comparison sign.

    (b) Two separate inequality statements joined by "and" or "or."

(c) A system of equations that includes at least one inequality.

(d) A solution set that is always expressed in interval notation.

2. When solving a compound inequality of the form
$$L \leq Ax + B < U,$$
which step should you perform first when simplifying the inequality?

(a) Divide every part by A.
(b) Subtract B from every part.
(c) Multiply every part by A.
(d) Add a constant to every part to balance the inequality.

3. When the coefficient of x in a compound inequality is negative, what must you do when dividing both sides by that coefficient?

(a) Leave the inequality symbols unchanged.
(b) Multiply each segment by -1 without changing the direction of the symbols.
(c) Reverse the inequality symbols to maintain the proper order.
(d) Square the coefficient first before proceeding.

4. Which of the following best defines the absolute value of a number x?

(a) The distance of x from zero on the number line.
(b) x itself if x is positive and 0 if x is negative.
(c) The result of multiplying x by -1 regardless of its sign.
(d) The square root of x squared, which may be negative.

5. Which of the following absolute value equations has no solution?

(a) $|x - 2| = 3$
(b) $|2x + 1| = -4$
(c) $|3x - 5| = 0$

(d) $|x+4| = 4$

6. When solving an equation of the form
$$|Ax + B| = C,$$
where C is a nonnegative constant, which two cases must be considered?

   (a) $Ax + B = C$ and $Ax + B = -C$
   (b) $Ax + B = C$ and $Ax + B = 0$
   (c) $Ax + B = 0$ and $Ax + B = -C$
   (d) $Ax + B = C + 1$ and $Ax + B = -C - 1$

7. In the worked example for the compound inequality
$$-2 \leq 3x - 1 < 8,$$
after adding 1 to each segment, what is the resulting inequality?

   (a) $-1 \leq 3x < 9$
   (b) $-3 \leq 3x < 7$
   (c) $-2 \leq 3x < 8$
   (d) $-1 \leq 3x \leq 9$

Answers:

1. **B: Two separate inequality statements joined by "and" or "or."**
   A compound inequality connects two distinct inequality statements using either "and" or "or" to express that a value must satisfy both (or at least one) of the conditions.

2. **B: Subtract B from every part.**
   The first step in isolating the variable is to eliminate the constant term B by subtracting it from every part of the inequality, thereby maintaining the balance across all segments.

3. **C: Reverse the inequality symbols to maintain the proper order.**
   When dividing by a negative coefficient, the direction of each inequality sign must be reversed to preserve the correct relational order among the expressions.

4. **A: The distance of x from zero on the number line.**
   The absolute value measures the distance of a number from zero, which is always nonnegative regardless of whether the number is positive or negative.

5. **B: $|2x+1| = -4$**
   Since absolute value represents the distance from zero, it can never yield a negative result. Therefore, an equation that sets an absolute value equal to a negative number has no valid solutions.

6. **A: $Ax + B = C$ and $Ax + B = -C$**
   When solving an absolute value equation of the form $|Ax + B| = C$ (with $C \geq 0$), you must split the problem into two cases—one where the expression inside the absolute value is C and one where it is $-C$.

7. **A: $-1 \leq 3x < 9$**
   Adding 1 to each part of the inequality $-2 \leq 3x - 1 < 8$ gives $-2 + 1 \leq 3x < 8 + 1$, simplifying to $-1 \leq 3x < 9$, which correctly represents the modified inequality.

Practice Problems

1. Solve the compound inequality:
$$-3 \leq 2x + 1 < 7$$

2. Solve the compound inequality:
$$1 \leq -3x + 4 < 10$$

3. Solve the absolute value equation:
$$|4x - 8| = 12$$

4. Solve the absolute value equation:
$$|2x + 5| = 3$$

5. Solve the absolute value equation:
$$|x - 3| = -2$$

6. Solve the absolute value equation:
$$|2x+5| = |x-3|$$

Answers

1. **Problem 1:** Solve
$$-3 \leq 2x + 1 < 7.$$

    Solution: We start by isolating the term containing x. Subtract 1 from each part of the inequality:
    $$-3 - 1 \leq 2x + 1 - 1 < 7 - 1,$$
    which simplifies to:
    $$-4 \leq 2x < 6.$$
    Next, divide each part by 2 (since 2 is positive, the inequality signs remain the same):
    $$\frac{-4}{2} \leq x < \frac{6}{2},$$
    giving:
    $$-2 \leq x < 3.$$
    Therefore, the solution set is all x such that $-2 \leq x < 3$.

2. **Problem 2:** Solve
$$1 \leq -3x + 4 < 10.$$

Solution: Begin by subtracting 4 from all three parts:
$$1 - 4 \leq -3x + 4 - 4 < 10 - 4,$$
which simplifies to:
$$-3 \leq -3x < 6.$$
Now, divide every part by -3. Remember, dividing by a negative number reverses the inequality signs:
$$\frac{-3}{-3} \geq x > \frac{6}{-3}.$$
This results in:
$$1 \geq x > -2,$$
or written in standard order:
$$-2 < x \leq 1.$$
Thus, x satisfies $-2 < x \leq 1$.

3. **Problem 3:** Solve the equation:
$$|4x - 8| = 12.$$

Solution: An absolute value equation $|A| = C$ (with $C \geq 0$) leads to two cases:

<u>Case 1:</u>
$$4x - 8 = 12.$$
Add 8 to both sides:
$$4x = 20,$$
then divide by 4:
$$x = 5.$$

<u>Case 2:</u>
$$4x - 8 = -12.$$
Add 8 to both sides:
$$4x = -4,$$
and divide by 4:
$$x = -1.$$

Therefore, the solutions are $x = 5$ and $x = -1$.

4. **Problem 4:** Solve the equation:
$$|2x + 5| = 3.$$

   Solution: We split the equation into two cases:
   Case 1:
   $$2x + 5 = 3.$$
   Subtract 5 from both sides:
   $$2x = -2,$$
   and divide by 2:
   $$x = -1.$$
   Case 2:
   $$2x + 5 = -3.$$
   Subtract 5 from both sides:
   $$2x = -8,$$
   and divide by 2:
   $$x = -4.$$
   Thus, the solutions are $x = -1$ and $x = -4$.

5. **Problem 5:** Solve the equation:
$$|x - 3| = -2.$$

   Solution: The definition of absolute value guarantees that $|x - 3| \geq 0$ for every real number x. Since -2 is negative, there is no real number x for which the absolute value equals -2. Therefore, the equation has **no solution**.

6. **Problem 6:** Solve the equation:
$$|2x + 5| = |x - 3|.$$

   Solution: When two absolute values are equal, $|A| = |B|$, it implies that either $A = B$ or $A = -B$. Accordingly, we consider two cases:
   Case 1:
   $$2x + 5 = x - 3.$$

Subtract x from both sides:
$$x + 5 = -3.$$
Then subtract 5 from each side:
$$x = -8.$$

Case 2:
$$2x + 5 = -(x - 3).$$
First, distribute the negative sign on the right:
$$2x + 5 = -x + 3.$$
Add x to both sides:
$$3x + 5 = 3.$$
Subtract 5 from both sides:
$$3x = -2,$$
and divide by 3:
$$x = -\frac{2}{3}.$$
Therefore, the solutions to the equation are $x = -8$ and $x = -\frac{2}{3}$.

Chapter 15

Graphing on the Cartesian Plane

The Cartesian Coordinate System

The Cartesian coordinate system serves as the fundamental framework for representing geometric and algebraic data visually. It is established by two perpendicular lines known as the x-axis and the y-axis. These axes intersect at a single point called the origin, denoted by (0, 0). The conceptual structure of this system divides the entire plane into four distinct regions, commonly referred to as the quadrants. In the first quadrant, both the x- and y-coordinates are positive; in the second quadrant, the x-coordinate is negative while the y-coordinate remains positive; in the third quadrant, both coordinates are negative; and in the fourth quadrant, the x-coordinate is positive while the y-coordinate is negative. Every point on the plane is represented by an ordered pair (x, y), where the first number indicates the horizontal displacement (positive numbers moving to the right and negative numbers to the left) and the second number indicates the vertical displacement (positive numbers moving upward and negative numbers downward). This systematic arrangement provides a clear relationship between numerical values and spatial locations, forming the basis for the geometric interpretation of algebraic expressions.

Plotting Points on the Cartesian Plane

Plotting points on the Cartesian plane entails marking locations that correspond to ordered pairs (x, y) accurately within the grid defined by the coordinate axes. The process begins with determining the x-coordinate, which specifies the horizontal position relative to the origin. A positive x value indicates a shift to the right, while a negative value indicates a shift to the left. Next, the y-coordinate is determined; it specifies the vertical displacement from the origin, with positive values signifying an upward movement and negative values a downward movement. For example, consider the point (3, -2). To plot this point, first locate 3 units to the right on the x-axis. From this position, move 2 units downward along the y-axis. Marking the point at the intersection of these determined positions results in an accurate representation of the coordinates provided by the ordered pair. The use of graph paper with uniformly spaced intervals enhances precision and facilitates the clear depiction of these points.

Geometric Interpretation of Algebraic Data

Graphing transforms algebraic expressions into a tangible geometric form, allowing for the visualization of relationships between variables. When algebraic data are represented as points on the Cartesian plane, they form distinct patterns that often reveal important characteristics of the underlying relationship. For instance, a series of points that align in a straight pattern suggests a linear relationship, which is typically represented by an equation such as

$$y = mx + b,$$

where m represents the slope and b the y-intercept. Through this representation, the slope conveys the rate of change between the variables, while the intercept shows where the line crosses the vertical axis. Furthermore, the spatial configuration of points can demonstrate symmetry, periodicity, and variation in rate, thereby aiding in the comprehension of more complex algebraic relationships. The geometric framework provided by the Cartesian plane enables the transformation of numeric or functional relationships into visual forms that are intuitive and informative. In this manner, the graphical representation of algebraic data not only facilitates the analysis of individual points but also serves as a powerful tool for interpreting trends and patterns inherent in mathematical models.

Multiple Choice Questions

1. Which of the following best defines the origin in the Cartesian coordinate system?

    (a) (1, 1)

    (b) (0, 0)

    (c) (-1, -1)

    (d) (0, 1)

2. In the Cartesian coordinate system, which quadrant contains points with a negative x-coordinate and a positive y-coordinate?

    (a) Quadrant I

    (b) Quadrant II

    (c) Quadrant III

    (d) Quadrant IV

3. To plot the point (3, -2) on the Cartesian plane, which sequence of steps is correct?

    (a) Move 3 units left along the x-axis, then 2 units up along the y-axis.

    (b) Move 3 units right along the x-axis, then 2 units down along the y-axis.

    (c) Move 3 units right along the x-axis, then 2 units up along the y-axis.

    (d) Move 3 units left along the x-axis, then 2 units down along the y-axis.

4. Which of the following accurately describes the ordered pair (x, y) in the Cartesian coordinate system?

    (a) x indicates vertical displacement; y indicates horizontal displacement.

    (b) x indicates horizontal displacement; y indicates vertical displacement.

(c) Both x and y indicate horizontal displacements.

(d) Both x and y indicate vertical displacements.

5. A series of points that align to form a straight line on a graph most likely represents which type of relationship?

   (a) Quadratic relationship.

   (b) Linear relationship.

   (c) Exponential relationship.

   (d) Radical relationship.

6. In the linear equation $y = mx + b$, what role does the parameter m play?

   (a) It represents the y-intercept.

   (b) It represents the slope of the line.

   (c) It represents the x-intercept.

   (d) It represents the origin.

7. How does graphing algebraic expressions on the Cartesian plane assist in understanding relationships between variables?

   (a) It transforms numerical data into a visual form that highlights trends, patterns, and symmetry.

   (b) It obscures the relationships by adding unnecessary complexity.

   (c) It provides exact numerical solutions without further interpretation.

   (d) It converts graphs into a table of values only.

Answers:

1. **B: (0, 0)**
   The origin is the intersection of the x-axis and y-axis, which is always at (0, 0).

2. **B: Quadrant II**
   In Quadrant II, the x-coordinate is negative while the y-coordinate is positive.

3. **B: Move 3 units right along the x-axis, then 2 units down along the y-axis.**
   Since 3 is positive, move right on the x-axis, and since -2 is negative, move down on the y-axis, which correctly locates the point (3, -2).

4. **B: x indicates horizontal displacement; y indicates vertical displacement.**
   In an ordered pair (x, y), the first coordinate represents horizontal displacement and the second represents vertical displacement.

5. **B: Linear relationship.**
   A straight line is the hallmark of a linear relationship, indicating a constant rate of change between the variables.

6. **B: It represents the slope of the line.**
   In the equation $y = mx + b$, m denotes the slope, which measures how steep the line is (i.e., the rate at which y changes with x).

7. **A: It transforms numerical data into a visual form that highlights trends, patterns, and symmetry.**
   Graphing on the Cartesian plane provides a visual framework that makes it easier to identify patterns, trends, and other key relationships between variables.

Practice Problems

1. Consider the point
$$(-3, 4).$$
Identify the quadrant in which this point is located.

2. Describe in detail the steps required to plot the point
$$(3, -2)$$
on the Cartesian plane.

3. Given the linear equation
$$y = 2x + 1,$$
identify its y-intercept and explain the importance of the y-intercept in graphing linear functions.

4. For the line represented by
$$y = -3x + 2,$$
determine the coordinates of the y-intercept and then find another distinct point on the line by choosing a suitable value for x.

5. Explain the significance of the origin, represented by

$$(0, 0),$$

on the Cartesian plane and discuss its role when graphing algebraic data.

6. Discuss how graphing algebraic data on the Cartesian plane aids in recognizing linear relationships. In your answer, address the roles of slope and intercept.

Answers

1. **Solution:** The point $(-3, 4)$ has a negative x-coordinate and a positive y-coordinate. On the Cartesian plane, a negative x indicates a position to the left of the origin, while a positive y indicates a location above the origin. This combination places the point in the **second quadrant**.

2. **Solution:** To plot the point $(3, -2)$ on the Cartesian plane, follow these steps:

    (a) Begin at the origin $(0, 0)$.

(b) Move along the x-axis: since the x-coordinate is 3, move 3 units to the right.

(c) From that position, move parallel to the y-axis: because the y-coordinate is -2, move 2 units downward.

(d) Mark the point where these movements intersect; this is the location of $(3, -2)$.

These steps ensure that the correct location is identified based on the horizontal and vertical displacements given in the ordered pair.

3. **Solution:** The equation
$$y = 2x + 1$$
is in slope-intercept form, $y = mx + b$, where m is the slope and b is the y-intercept. To find the y-intercept, set $x = 0$:
$$y = 2(0) + 1 = 1.$$
Therefore, the y-intercept is
$$(0, 1).$$
The y-intercept is important because it represents the point where the line crosses the y-axis. It serves as a starting reference for graphing the line and, when combined with the slope, helps in determining the entire line's behavior.

4. **Solution:** For the line
$$y = -3x + 2,$$
the y-intercept is found by setting $x = 0$:
$$y = -3(0) + 2 = 2.$$
Thus, the y-intercept is
$$(0, 2).$$
To find another point on the line, choose a convenient value for x, for example, $x = 1$:
$$y = -3(1) + 2 = -3 + 2 = -1.$$
So another point on the line is
$$(1, -1).$$
These two points are sufficient to graph the line accurately.

5. **Solution:** The origin, $(0,0)$, is the point where the x-axis and y-axis intersect. It represents the starting point or the zero point for both axes. In graphing algebraic data, the origin serves as a reference point from which every other point's position is measured. Its role is crucial because it provides a common baseline for comparing the relative positions of points and for recognizing patterns in the data when plotted on the Cartesian plane.

6. **Solution:** Graphing algebraic data on the Cartesian plane helps in visually identifying linear relationships. When points that represent algebraic expressions form a straight-line pattern, it indicates a linear relationship with a constant rate of change. The slope of the line, calculated as the ratio of the change in y to the change in x, shows how steep the line is and thus how much y changes for a given change in x. The y-intercept indicates the point where the line crosses the y-axis, offering insight into the starting value of the relationship when $x = 0$. Together, the slope and y-intercept provide a complete description of the linear relationship, making it easier to predict values and understand the behavior of the data.

Chapter 16

Understanding Slope and Intercepts

The Concept of Slope

In any linear equation, the slope indicates the magnitude and direction of change between two variables along a straight line. The slope is defined as the ratio of the vertical change, known as the rise, to the horizontal change, known as the run. Mathematically, the slope (m) can be expressed as

$$m = \frac{\Delta y}{\Delta x} = \frac{y_2 - y_1}{x_2 - x_1}.$$

A positive value for m signifies that as the horizontal variable increases, the vertical variable also increases, generating an upward trend when moving from left to right. Conversely, a negative value for m indicates that the vertical variable decreases as the horizontal variable increases, resulting in a downward trend. Special cases include a zero slope, corresponding to a horizontal line, and an undefined slope, which characterizes a vertical line. These properties play a central role in the study and analysis of linear relationships.

The y-Intercept: Identifying the Vertical Intersection

The y-intercept is the point where the graph of a linear equation crosses the y-axis. By definition, at the y-intercept the value of x is zero. In the slope-intercept form of a linear equation, written as

$$y = mx + b,$$

the constant b represents the y-intercept. This single value imparts a clear starting point for graphing the line by indicating the initial value of y when no horizontal displacement has occurred. The precise determination of the y-intercept is crucial since it anchors the position of the line on the Cartesian plane. Understanding the y-intercept facilitates the visualization of the entire linear relationship by coupling it with the slope to predict the behavior of the line as x varies.

The x-Intercept: Identifying the Horizontal Intersection

The x-intercept marks the point at which the line crosses the x-axis, where the value of y is zero. To determine the x-intercept for a linear equation, the value $y = 0$ is substituted into the equation, and the resulting equation is solved for x. For example, given the equation

$$y = mx + b,$$

setting y equal to zero yields

$$0 = mx + b,$$

which can then be rearranged to isolate x:

$$x = -\frac{b}{m},$$

provided that the slope m is not zero. The x-intercept provides critical information about the horizontal placement of the line and serves as a reference point for comprehending the overall behavior of the linear function. Its role is particularly significant when assessing the point at which the output of the function reaches zero.

Analyzing Linear Relationships Through Slope and Intercepts

The properties of slope, y-intercept, and x-intercept together offer a comprehensive description of a linear function. The slope quantifies the constant rate at which the vertical variable changes with respect to the horizontal variable, effectively measuring the steepness of the line and indicating the direction of the trend. Simultaneously, the y-intercept and x-intercept establish the precise points at which the line crosses the coordinate axes, thereby anchoring the line to specific locations on the graph. In the standard linear equation form, these elements interplay to form a complete model of the relationship between variables.

The explicit characterization of the slope along with the intercepts allows for the systematic determination of any point along the line. Moreover, the graphical representation of these parameters provides clarity in predicting how variations in one variable influence the other. This detailed understanding lays the groundwork for more advanced applications in algebra and serves as a robust framework for the analysis of linear relationships in various contexts.

Multiple Choice Questions

1. Which of the following formulas correctly defines the slope (m) of a line passing through two points (x_1, y_1) and (x_2, y_2) where $x_2 \neq x_1$?

    (a) $m = \frac{x_2 - x_1}{y_2 - y_1}$

    (b) $m = \frac{y_2 - y_1}{x_2 - x_1}$

    (c) $m = \frac{x_1 - x_2}{y_2 - y_1}$

    (d) $m = \frac{y_1 - y_2}{x_1 - x_2}$

2. In the slope-intercept equation $y = mx + b$, which statement about the constant b is correct?

    (a) b represents the slope of the line.

    (b) b represents the y-intercept of the line.

    (c) b represents the x-intercept of the line.

(d) b represents the coefficient of x.

3. A line with a zero slope is best described as:

    (a) Vertical.
    (b) Horizontal.
    (c) Diagonal with a positive incline.
    (d) Diagonal with a negative incline.

4. To find the x-intercept of the line given by $y = mx + b$ (with $m \neq 0$), you should:

    (a) Set $x = 0$ and solve for y.
    (b) Set $y = 0$ and solve for x.
    (c) Set $m = 0$ and solve for x.
    (d) Set $b = 0$ and solve for y.

5. If a line has a negative slope, this implies that as x increases:

    (a) y increases.
    (b) y decreases.
    (c) y remains constant.
    (d) The line becomes vertical.

6. An undefined slope in a linear equation most typically corresponds to a:

    (a) Horizontal line.
    (b) Vertical line.
    (c) Line with a zero y-intercept.
    (d) Diagonal line with inconsistent change.

7. Why is the y-intercept considered a crucial component when graphing a linear equation in slope-intercept form?

    (a) It determines the steepness of the line.
    (b) It provides the starting value of y when $x = 0$.
    (c) It indicates the rate of change between x and y.
    (d) It identifies where the line crosses the x-axis.

Answers:

1. **B:** $m = \frac{y_2 - y_1}{x_2 - x_1}$
   This formula correctly defines the slope by calculating the ratio of the vertical change (rise) to the horizontal change (run) between two distinct points.

2. **B: b represents the y-intercept of the line**
   In the equation $y = mx + b$, b is the constant that gives the point where the line crosses the y-axis (i.e., when $x = 0$).

3. **B: Horizontal**
   A zero slope means there is no vertical change as x varies, which characterizes a horizontal line.

4. **B: Set $y = 0$ and solve for x**
   To find the x-intercept, set y equal to zero in the equation and solve for x, thereby determining where the line crosses the x-axis.

5. **B: y decreases**
   A negative slope indicates that as x increases, the value of y decreases, meaning the line falls as it moves from left to right.

6. **B: Vertical line**
   An undefined slope occurs when the denominator (change in x) is zero, which happens in a vertical line where all the points share the same x-value.

7. **B: It provides the starting value of y when $x = 0$**
   The y-intercept is critical because it anchors the graph of the line by indicating the value of y when there is no horizontal displacement, serving as a reference point for plotting the rest of the line.

Practice Problems

1. Compute the slope of the line passing through the points (1, 2) and (4, 10). Use the slope formula

$$m = \frac{y_2 - y_1}{x_2 - x_1}.$$

2. For the linear equation
$$y = -3x + 7,$$
identify the y-intercept and explain its significance on the graph.

3. For the linear equation
$$y = \frac{1}{2}x - 4,$$
determine the x-intercept. Show your work by setting y equal to 0 and solving for x.

4. Write the equation of the line that has a slope of 4 and a y-intercept of -3. Express your answer in slope-intercept form and explain what each component of the equation represents.

5. Convert the equation
$$2y - 6x = 8$$
into slope-intercept form. Then, identify the slope and the y-intercept, providing a step-by-step explanation of your process.

6. A line has a slope of $-\frac{2}{3}$ and passes through the point $(3, 1)$. Find the equation of the line in slope-intercept form using the point-slope method, and explain each step in your calculation.

Answers

1. **Solution:** To find the slope, we use the formula
$$m = \frac{y_2 - y_1}{x_2 - x_1}.$$
Here, let $(x_1, y_1) = (1, 2)$ and $(x_2, y_2) = (4, 10)$. Substituting these values gives:
$$m = \frac{10 - 2}{4 - 1} = \frac{8}{3}.$$

Therefore, the slope of the line is $\frac{8}{3}$. This means that for every increase of 3 units in the x-direction, the y-value increases by 8 units.

2. **Solution:** The equation given is
$$y = -3x + 7.$$
In slope-intercept form, $y = mx + b$, the y-intercept is represented by the constant term b. Here, $b = 7$, so the y-intercept is the point $(0, 7)$. This point indicates where the line crosses the y-axis and serves as the starting value of y when $x = 0$.

3. **Solution:** The equation provided is
$$y = \frac{1}{2}x - 4.$$
To find the x-intercept, we set $y = 0$:
$$0 = \frac{1}{2}x - 4.$$
Adding 4 to both sides yields:
$$\frac{1}{2}x = 4.$$
Multiplying both sides by 2 gives:
$$x = 8.$$
Thus, the x-intercept is $(8, 0)$, which represents the point where the line crosses the x-axis.

4. **Solution:** A line with a slope of 4 and a y-intercept of -3 is written in slope-intercept form as:
$$y = 4x - 3.$$
In this equation, 4 is the slope (indicating that for every one unit increase in x, y increases by 4 units), and -3 is the y-intercept (indicating that the line crosses the y-axis at the point $(0, -3)$).

5. **Solution:** We start with the equation:
$$2y - 6x = 8.$$

The goal is to convert this to slope-intercept form, $y = mx+b$. First, add $6x$ to both sides:

$$2y = 6x + 8.$$

Next, divide every term by 2 to isolate y:

$$y = 3x + 4.$$

Here, the slope m is 3 and the y-intercept b is 4 (so the line crosses the y-axis at $(0, 4)$).

6. **Solution:** We are given the slope $m = -\frac{2}{3}$ and a point $(3, 1)$ through which the line passes. Using the point-slope form:

$$y - y_1 = m(x - x_1),$$

we substitute $x_1 = 3$, $y_1 = 1$, and $m = -\frac{2}{3}$:

$$y - 1 = -\frac{2}{3}(x - 3).$$

Distribute the slope on the right-hand side:

$$y - 1 = -\frac{2}{3}x + 2.$$

Finally, add 1 to both sides to solve for y:

$$y = -\frac{2}{3}x + 3.$$

Thus, the equation of the line in slope-intercept form is

$$y = -\frac{2}{3}x + 3.$$

This shows that the line has a slope of $-\frac{2}{3}$ and crosses the y-axis at $(0, 3)$.

Chapter 17

Graphing Linear Functions

Graphing with Slope-Intercept Form

A linear function expressed in the slope-intercept form appears as

$$y = mx + b.$$

In this expression, the coefficient m represents the slope of the line, quantifying the constant ratio of the vertical change to the horizontal change between any two points. The constant b denotes the y-intercept, which is the point on the Cartesian plane where the line crosses the y-axis. At this intersection, the x-coordinate is zero, and b provides an immediate reference for the vertical position of the line. Once the y-intercept is identified, the slope m determines the increments by which subsequent points on the line are derived through the relationship of rise over run. A positive value of m generates an ascending line, while a negative value produces a descending line. The fractional form of the slope, when reduced, aids in systematically marking equal intervals along the horizontal axis and moving proportionally in the vertical direction to render a precise straight-line graph.

Graphing with Point-Slope Form

An alternative method for graphing a linear function is provided by the point-slope form, given as

$$y - y_1 = m(x - x_1).$$

In this formulation, the coordinates (x_1, y_1) indicate a specific point through which the line passes, and m continues to signify the slope. The established point acts as an anchor on the graph, from which the slope imposes a consistent direction and magnitude of change. The representation explicitly emphasizes the difference between any arbitrary point (x, y) on the line and the designated point (x_1, y_1). Deriving additional points involves applying the constant rate of change m over the horizontal displacement from x_1. This method proves particularly effective when the coordinates of a known point are available, as it directly integrates that point into the equation. The point-slope form provides a clear framework that encapsulates both the positional information and the directional trend of the line, thereby facilitating an accurate and efficient plotting of the linear function.

Interpreting the Behavior of Linear Functions

The straight line produced by a linear function encapsulates a constant rate of change, a feature that is inherently defined by the slope and the intercept. The slope determines the steepness and the directional flow of the line: a larger absolute value of m indicates a steeper line, whereas a smaller absolute value corresponds to a gentler incline. The sign of m is critical; a positive slope consistently yields an upward progression from left to right, while a negative slope results in a downward progression. In conjunction with the slope, the y-intercept provides a fixed point on the graph where the function initiates its vertical trajectory. This combination of parameters ensures that every point on the line adheres to an immutable proportional relationship between the variables. The uniformity of this relationship is reflected in the linear appearance of the graph and serves as an essential characteristic when analyzing the behavior of linear functions.

Multiple Choice Questions

1. Which form of a linear equation directly displays the y-intercept?

    (a) Point-Slope Form

    (b) Slope-Intercept Form

    (c) Standard Form

    (d) Factored Form

2. In the slope-intercept equation
$$y = mx + b,$$
what does the coefficient m represent?

    (a) The y-intercept where the line crosses the y-axis

    (b) The slope of the line, indicating the ratio of vertical change to horizontal change

    (c) The x-intercept where the line crosses the x-axis

    (d) A fixed point on the line used as a reference

3. In the point-slope form
$$y - y_1 = m(x - x_1),$$
what does the coordinate pair (x_1, y_1) denote?

    (a) The slope of the line

    (b) The y-intercept of the line

    (c) A specific point on the line used as an anchor for graphing

    (d) The midpoint of the line segment

4. What is the first step when graphing a linear equation given in slope-intercept form?

    (a) Identify and plot the x-intercept

    (b) Identify and plot the y-intercept

    (c) Convert the equation to standard form

    (d) Determine the vertex of the line

5. How does a positive slope affect the graph of a linear function?

    (a) The line ascends from left to right
    (b) The line descends from left to right
    (c) The line is horizontal
    (d) The line is vertical

6. When using the point-slope form $y - y_1 = m(x - x_1)$ to graph a line, which of the following procedures is correct?

    (a) Plot the point (x_1, y_1), then use the slope m to determine a second point by moving "rise over run"
    (b) Plot the y-intercept first, then adjust using the slope regardless of the given point
    (c) Determine two arbitrary points using the slope and connect them
    (d) Solve for y to convert to slope-intercept form before plotting

7. Consider the equation
$$y - 3 = -2(x - 1).$$
Which of the following statements is true about its graph?

    (a) The line passes through $(3, -2)$ with a slope of $\frac{1}{2}$
    (b) The line passes through $(1, 3)$ with a slope of -2
    (c) The line has a y-intercept of 3 and a slope of -2
    (d) The line passes through $(-1, -3)$ with a slope of 2

Answers:

1. **B: Slope-Intercept Form**
   This form, given by $y = mx + b$, directly shows the y-intercept b, which is the starting point when graphing the line.

2. **B: The slope of the line, indicating the ratio of vertical change to horizontal change**
   In the equation $y = mx + b$, the coefficient m determines how steep the line is, representing the change in y for each unit change in x.

3. **C: A specific point on the line used as an anchor for graphing**
   The point (x_1, y_1) in the point-slope form is a known point on the line, which, along with the slope m, is used to graph the line.

4. **B: Identify and plot the y-intercept**
   When graphing from the slope-intercept form, the y-intercept b is the easiest point to plot first since it directly gives the intersection with the y-axis.

5. **A: The line ascends from left to right**
   A positive slope means that as x increases, y also increases, resulting in an upward trend from left to right.

6. **A: Plot the point (x_1, y_1), then use the slope m to determine a second point by moving "rise over run"**
   The point-slope form provides an anchor point and the slope. After plotting (x_1, y_1), applying the slope as a ratio (rise/run) yields another point for drawing the line.

7. **B: The line passes through $(1, 3)$ with a slope of -2**
   In the equation $y - 3 = -2(x - 1)$, the point $(1, 3)$ is used as the reference point, and the slope of the line is -2, indicating that for each unit increase in x, y decreases by 2.

Practice Problems

1. Graph the linear function
$$y = 2x - 3.$$
Identify the slope and y-intercept. Then, starting with the y-intercept, use the slope to plot at least two additional points on the Cartesian plane and sketch the line.

2. Write the equation of a line that passes through the point $(3, 4)$ with a slope of -5. First, express the equation in point-slope form, then convert it to slope-intercept form. Finally, sketch the graph of the line.

3. For the linear function
$$y = -\frac{1}{3}x + 2,$$
start at the y-intercept and use the slope to determine two additional points on the graph. Explain your process clearly.

4. A cell phone plan charges a fixed monthly fee of 30 and an additional fee of 0.10 per minute of usage. Represent this plan with a linear function in slope-intercept form where the number of minutes (m) is the independent variable. Identify the slope and y-intercept and interpret their meaning in context. Then, sketch the graph of the function.

5. Given the two points $(2, -1)$ and $(6, 7)$, first determine the slope of the line passing through them. Then, use the point-slope form to write the equation of the line and convert it into slope-intercept form. Provide a detailed explanation of each step.

6. Consider the general linear equation
$$y = mx + b.$$
Explain how varying the values of m (the slope) and b (the y-intercept) affects the graph of the line. Provide specific examples (for instance, compare $m = 2$, $b = 1$ with $m = -2$, $b = -1$) to illustrate your explanation.

Answers

1. **Solution:** The given equation is
$$y = 2x - 3.$$
Here, the slope $m = 2$ and the y-intercept $b = -3$. This tells us that the line crosses the y-axis at the point $(0, -3)$. To graph, start at the y-intercept $(0, -3)$. The slope of 2 means that for every 1 unit increase in x, y increases by 2 units.

- From $(0, -3)$, moving one unit to the right ($x = 1$) will increase y by 2 to give the point $(1, -1)$.
- From $(1, -1)$, moving one unit to the right again ($x = 2$) gives the point $(2, 1)$.

Plot these points and draw a straight line through them.

2. **Solution:** Given the point $(3, 4)$ and slope $m = -5$, use the point-slope form:

$$y - y_1 = m(x - x_1).$$

Substitute $(x_1, y_1) = (3, 4)$ and $m = -5$:

$$y - 4 = -5(x - 3).$$

To convert to slope-intercept form, expand the right side:

$$y - 4 = -5x + 15.$$

Then add 4 to both sides:

$$y = -5x + 19.$$

This gives the slope-intercept form $y = -5x + 19$. To graph, plot the point $(3, 4)$ and use the slope -5 (meaning a drop of 5 units for every 1 unit increase in x) to plot additional points.

3. **Solution:** The equation is

$$y = -\frac{1}{3}x + 2.$$

The y-intercept is at $(0, 2)$. The slope $-\frac{1}{3}$ indicates that for every increase of 3 in x, y decreases by 1 unit (or equivalently, for every 1 unit increase in x, y decreases by $\frac{1}{3}$).

- Starting at $(0, 2)$, if you increase x by 3 (to $x = 3$), then y decreases by 1, giving the point $(3, 1)$.
- Similarly, if you decrease x by 3 (to $x = -3$), then y increases by 1, resulting in the point $(-3, 3)$.

These steps allow you to mark additional points and accurately sketch the graph.

4. **Solution:** Let m represent the number of minutes. The cost function is expressed as:
$$C(m) = 0.10m + 30.$$
Here, the slope 0.10 represents the additional charge per minute (i.e., the variable part of the cost), and the y-intercept 30 represents the fixed monthly fee. This function implies that even if no minutes are used ($m = 0$), the cost is 30. To graph, plot the point $(0, 30)$ and use the slope 0.10 (a rise of 0.10 for every 1 unit increase in m) to plot following points.

5. **Solution:** First, compute the slope using the points $(2, -1)$ and $(6, 7)$:
$$m = \frac{7 - (-1)}{6 - 2} = \frac{8}{4} = 2.$$
Next, use the point-slope form with one of the points (using $(2, -1)$ for example):
$$y - (-1) = 2(x - 2),$$
which simplifies to:
$$y + 1 = 2(x - 2).$$
Expanding the right side:
$$y + 1 = 2x - 4.$$
Then subtract 1 from both sides to convert to slope-intercept form:
$$y = 2x - 5.$$
This step-by-step process (calculating the slope, writing the point-slope form, and converting to slope-intercept form) yields the final equation of the line.

6. **Solution:** In the general linear equation
$$y = mx + b,$$
the parameter m (slope) determines the steepness and direction of the line. A positive m causes the line to rise from left to right, whereas a negative m causes it to fall. The parameter b (y-intercept) specifies the point where the line crosses the y-axis.

Example 1: If $m = 2$ and $b = 1$, the equation is

$$y = 2x + 1.$$

This line is steeply rising and intersects the y-axis at $(0, 1)$.

Example 2: If $m = -2$ and $b = -1$, then the equation is

$$y = -2x - 1.$$

Here, the line is steeply falling and crosses the y-axis at $(0, -1)$.

In summary, increasing the magnitude of m makes the line steeper, while the sign of m (positive or negative) determines its direction. Changing b shifts the entire line up or down without altering its slope.

Chapter 18

Graphing Linear Inequalities

Understanding Linear Inequalities

A linear inequality resembles a linear equation but replaces the equality symbol with one of the inequality symbols (<, >, , or). In this context, an inequality such as

$$ax + by \leq c$$

defines a set of ordered pairs (x, y) that satisfy the stated condition. Unlike a linear equation, which corresponds to a single line, a linear inequality represents a continuous region on the coordinate plane. The boundary of this region is determined by the corresponding linear equation, and the remainder of the graph indicates all points that satisfy the inequality condition. The distinction between strict inequalities (< or >) and inclusive inequalities (or) establishes the nature of the boundary line, as will be discussed in subsequent sections.

Plotting the Boundary Line

The boundary line is derived by replacing the inequality symbol with an equality symbol to form the equation

$$ax + by = c.$$

When graphing this equation, it is essential first to determine its intercepts; the x-intercept is obtained by setting $y = 0$ and solving for x, and the y-intercept is found by setting $x = 0$ and solving for y. Once both intercepts have been calculated, the boundary line can be drawn through these two points. In instances where a strict inequality is considered, the boundary line is rendered as a dashed line to signify that the points on the line are not included in the solution set. Conversely, if an inclusive inequality is present, the boundary line appears as a solid line, indicating that points on the line satisfy the inequality.

Determining the Solution Region

After graphing the boundary line, the next stage involves identifying the region of the coordinate plane that represents the solutions to the inequality. This is accomplished by selecting a test point that does not lie on the boundary line; the origin $(0,0)$ is frequently chosen if it is not coincident with the boundary. By substituting the coordinates of the test point into the original inequality, it can be verified whether the resulting statement is true. If the inequality holds for the test point, then all points in that region are part of the solution set, and the region is shaded accordingly. Should the test point fail to satisfy the inequality, the opposite region is identified as the solution set and subsequently shaded. This process ensures that the graphical representation of the inequality accurately reflects the set of all ordered pairs (x, y) that meet the specified condition.

Worked Example in Graphing Inequalities

Consider the linear inequality
$$2x - 3y > 6.$$

The boundary line is first determined by converting the inequality into the corresponding equality:
$$2x - 3y = 6.$$

In order to graph the boundary, the y-intercept is calculated by setting $x = 0$, yielding
$$-3y = 6 \implies y = -2.$$
Similarly, the x-intercept is found by setting $y = 0$, which results in
$$2x = 6 \implies x = 3.$$
With the intercepts $(0, -2)$ and $(3, 0)$ identified, the boundary line is drawn through these two points. Due to the strict inequality $(>)$ present in the original statement, the boundary line is drawn as a dashed line to indicate that points on this line are not included in the solution set.

To determine which side of the boundary line comprises the solution region, the test point $(0, 0)$ is substituted into the original inequality. The substitution yields
$$2(0) - 3(0) > 6 \implies 0 > 6,$$
a false statement. Consequently, the region that does not contain the origin represents the set of all solutions to the inequality. Once the appropriate region is identified, that side of the boundary line is shaded to illustrate the solution region on the coordinate plane.

Each step in this process—determining the boundary line, distinguishing between dashed and solid representations, and testing the region using a sample point—plays an integral role in visualizing the solution set of a linear inequality. The carefully executed graph not only identifies the set of all possible ordered pairs (x, y) satisfying the inequality but also emphasizes the continuous nature of the solution region on the coordinate plane.

Multiple Choice Questions

1. Which of the following best describes the difference between a strict inequality and an inclusive inequality when graphing?

    (a) Strict inequalities use a solid line while inclusive inequalities use a dashed line.

    (b) Strict inequalities use a dashed line while inclusive inequalities use a solid line.

    (c) Both strict and inclusive inequalities always use dashed lines.

(d) Both strict and inclusive inequalities always use solid lines.

2. When converting the inequality ax + by c to its corresponding boundary line for graphing, what is the correct procedure to determine the intercepts?

   (a) Set x = 0 to find the x-intercept and y = 0 to find the y-intercept.
   (b) Set y = 0 to find the x-intercept and x = 0 to find the y-intercept.
   (c) Set both x and y equal to c, then solve.
   (d) Rewrite the inequality in slope-intercept form to read the intercepts directly.

3. What is the primary purpose of using a test point when graphing a linear inequality?

   (a) To determine the slope of the boundary line.
   (b) To check if the intercepts were calculated correctly.
   (c) To decide which side of the boundary line should be shaded as the solution region.
   (d) To convert the inequality into its equivalent equation.

4. In the worked example with the inequality

$$2x - 3y > 6,$$

   why is the boundary line drawn as a dashed line?

   (a) Because the intercepts are not included.
   (b) Because the inequality is strict, meaning points on the boundary are not part of the solution.
   (c) Because the line represents only half of the solution set.
   (d) Because the inequality was first converted to a standard form.

5. Why is it beneficial to determine both the x-intercept and y-intercept when graphing the boundary line of a linear inequality?

   (a) It simplifies the process by providing two clear points through which the line can be accurately drawn.

(b) It automatically shades the correct solution region.

(c) It confirms whether the inequality symbol should be strict or inclusive.

(d) It enables you to calculate the slope of the inequality.

6. Which test point is most commonly used when deciding which region satisfies a linear inequality?

   (a) (1, 1)
   (b) (0, 0), provided it is not on the boundary line
   (c) (1, 0)
   (d) (-1, -1)

7. What does the boundary line represent in the graph of a linear inequality?

   (a) The entire solution set of the inequality.
   (b) The set of points where the original inequality holds true.
   (c) The set of points that satisfy the corresponding equality, ax + by = c.
   (d) The region to be shaded on the coordinate plane.

Answers:

1. **B: Strict inequalities use a dashed line while inclusive inequalities use a solid line.** This is because in strict inequalities (using < or >), points on the boundary line are not considered part of the solution set, so a dashed line is used. In contrast, inclusive inequalities (using or) include the boundary points, which is indicated by a solid line.

2. **B: Set y = 0 to find the x-intercept and x = 0 to find the y-intercept.** To graph the boundary line of ax + by c, replace the inequality with an equality. The x-intercept is found by setting y = 0 and solving for x, and the y-intercept is found by setting x = 0 and solving for y.

3. **C: To decide which side of the boundary line should be shaded as the solution region.** A test point (often (0,0) when it is not on the line) is substituted into the original inequality. If the inequality holds true for the test point, then that side of the line is the solution region.

4. **B: Because the inequality is strict, meaning points on the boundary are not part of the solution.** For strict inequalities (using > or <), the boundary line is drawn as a dashed line to visually indicate that the points on that line are not included in the solution set.

5. **A: It simplifies the process by providing two clear points through which the line can be accurately drawn.** Computing the x- and y-intercepts gives two distinct points on the boundary line, making it easier and more accurate to draw the line on the coordinate plane.

6. **B: (0, 0), provided it is not on the boundary line.** The origin (0,0) is typically chosen as the test point because it is simple to compute with, as long as it does not lie on the boundary line.

7. **C: The set of points that satisfy the corresponding equality, ax + by = c.** The boundary line is found by converting the inequality into an equality. It represents the dividing line between regions where the inequality holds true and where it does not, though the line itself is only part of the solution when the inequality is inclusive.

Practice Problems

1. Given the inequality

$$3x + 2y \leq 12,$$

a) Find the x-intercept and y-intercept of the boundary line.
b) State whether the boundary line should be drawn as solid or dashed.
c) Use the test point (0,0) to determine the correct solution region.

2. Given the inequality
$$-2x + y > 4,$$

a) Determine the x-intercept and y-intercept of the boundary line by converting to the equality
$-2x + y = 4$.
b) Explain whether the boundary line is dashed or solid.
c) Use the test point (0,0) to decide which half-plane represents the solution set.

3. Consider the inequality
$$\frac{x}{2} - y \geq -3.$$

a) Convert the inequality to the equality $\frac{x}{2} - y = -3$ and find its x-intercept and y-intercept.
b) Indicate the correct style (dashed or solid) for the boundary line.
c) Test the point (0,0) to determine which region satisfies the inequality.

4. Graph the inequality
$$4x - 5y < 10.$$

a) Find the x-intercept and y-intercept of the boundary line by solving $4x - 5y = 10$.
b) Determine whether the boundary line should be dashed or solid.
c) Use the test point (0,0) to verify the correct solution region.

5. For the inequality
$$2x + y \geq 4,$$
verify if the point (2, 0) satisfies the inequality. Provide your working steps.

6. Write the complete steps to graph the inequality
$$-x + 4y \leq 8.$$
Include:
a) Finding the x-intercept and y-intercept from the equation $-x + 4y = 8$.
b) Deciding whether the boundary line is dashed or solid.
c) Testing a point (use (0,0) if it is not on the boundary) to determine the correct solution region.

Answers

1. For the inequality
$$3x + 2y \le 12,$$
first write the corresponding boundary equation:
$$3x + 2y = 12.$$
To find the x-intercept, set $y = 0$:
$$3x = 12 \implies x = 4.$$
Thus, the x-intercept is $(4, 0)$.
To find the y-intercept, set $x = 0$:
$$2y = 12 \implies y = 6.$$
Thus, the y-intercept is $(0, 6)$.
Since the inequality is $\le$ (inclusive), the boundary line is drawn as a solid line.
Testing the point (0,0):
$$3(0) + 2(0) = 0 \le 12,$$
which is true. Therefore, the solution region is the half-plane that includes the point (0,0).

2. For the inequality
$$-2x + y > 4,$$
write the boundary equation:
$$-2x + y = 4.$$
To find the y-intercept, set $x = 0$:
$$y = 4 \implies (0, 4).$$
To find the x-intercept, set $y = 0$:
$$-2x = 4 \implies x = -2,$$
yielding the point $(-2, 0)$.
Since the inequality is strict ($>$) the boundary line is drawn

as a dashed line.
Testing the point (0,0):
$$-2(0) + 0 = 0,$$
and since $0 > 4$ is false, (0,0) is not in the solution region. Hence, the solution region is the half-plane not containing the origin.

3. For the inequality
$$\frac{x}{2} - y \geq -3,$$
first convert it into the equality:
$$\frac{x}{2} - y = -3.$$
To find the x-intercept, let $y = 0$:
$$\frac{x}{2} = -3 \implies x = -6,$$
so the x-intercept is $(-6, 0)$.
To find the y-intercept, let $x = 0$:
$$-y = -3 \implies y = 3,$$
so the y-intercept is $(0, 3)$.
Because the inequality is $\geq$ (inclusive), the boundary line is solid.
Testing (0,0):
$$\frac{0}{2} - 0 = 0,$$
and since $0 \geq -3$ is true, the region containing (0,0) is the solution set.

4. For the inequality
$$4x - 5y < 10,$$
write the boundary equation:
$$4x - 5y = 10.$$
To find the x-intercept, set $y = 0$:
$$4x = 10 \implies x = \frac{10}{4} = 2.5,$$

resulting in the point $(2.5, 0)$.
To find the y-intercept, set $x = 0$:
$$-5y = 10 \implies y = -2,$$
resulting in the point $(0, -2)$.
Since the inequality is strict ($<$), the boundary line is drawn as a dashed line.
Testing $(0,0)$:
$$4(0) - 5(0) = 0 < 10,$$
which is true; thus, the solution region is the half-plane that includes $(0,0)$.

5. For the inequality
$$2x + y \geq 4,$$
substitute the point $(2,0)$:
$$2(2) + 0 = 4.$$
Since $4 \geq 4$ is true, the point $(2,0)$ satisfies the inequality.

6. For the inequality
$$-x + 4y \leq 8,$$
first write the corresponding equality:
$$-x + 4y = 8.$$
To find the x-intercept, set $y = 0$:
$$-x = 8 \implies x = -8,$$
so the x-intercept is $(-8, 0)$.
To find the y-intercept, set $x = 0$:
$$4y = 8 \implies y = 2,$$
so the y-intercept is $(0, 2)$.
Since the inequality is $\leq$ (inclusive), the boundary line is drawn as a solid line.
Testing the point $(0,0)$:
$$-0 + 4(0) = 0 \leq 8,$$
which is true; therefore, the solution region is the half-plane that contains $(0,0)$.

Chapter 19

Solving Systems of Linear Equations

Overview of Systems of Equations

A system of linear equations consists of two or more equations that share common variables. The objective is to determine the values for these variables that satisfy every equation in the system simultaneously. In typical high school problems, many systems involve two equations with two unknowns, although the principles extend naturally to systems involving three or more variables. Methods such as substitution and elimination provide systematic techniques to reduce the system to a simpler form, making it possible to solve for one variable at a time.

Method of Substitution

The substitution method involves solving one equation for a chosen variable and then substituting the resulting expression into the other equation(s). This procedure reduces the number of variables, thereby transforming the original system into an equivalent system that is simpler to analyze. The process requires precise algebraic manipulation to ensure that the operations performed do not alter the intrinsic relationships between the variables.

1 Worked Example: Substitution Method

Consider the system of equations
$$x + 2y = 7,$$
$$3x - y = 5.$$
The first equation is solved for x by isolating the variable:
$$x = 7 - 2y.$$
This expression for x is substituted into the second equation, yielding
$$3(7 - 2y) - y = 5.$$
Expansion and simplification provide
$$21 - 6y - y = 5,$$
$$21 - 7y = 5.$$
Subtracting 21 from both sides results in
$$-7y = -16,$$
and dividing both sides by -7 gives
$$y = \frac{16}{7}.$$
Substituting the value of y back into the expression for x produces
$$x = 7 - 2\left(\frac{16}{7}\right) = 7 - \frac{32}{7} = \frac{49 - 32}{7} = \frac{17}{7}.$$
The solution $x = \frac{17}{7}$ and $y = \frac{16}{7}$ satisfies both equations, demonstrating the effectiveness of the substitution method.

Method of Elimination

The elimination method centers on the removal of one variable by the strategic addition or subtraction of the given equations. By multiplying one or both equations by appropriate constants, it is possible to align the coefficients of a selected variable so that they become equal in magnitude but opposite in sign. Adding or subtracting the equations then cancels the chosen variable, resulting in a simplified equation with one fewer variable. This technique can be applied iteratively to systems involving multiple variables.

1 Worked Example: Elimination Method

Examine the system of equations

$$2x + 3y = 8,$$

$$4x - y = 2.$$

The elimination method aims to remove the variable y. Multiplication of the second equation by 3 yields

$$3(4x - y) = 3 \cdot 2,$$

or equivalently,

$$12x - 3y = 6.$$

The original first equation remains unchanged:

$$2x + 3y = 8.$$

Adding the two equations eliminates y:

$$(2x + 3y) + (12x - 3y) = 8 + 6,$$

which simplifies to

$$14x = 14.$$

Dividing both sides by 14 determines the value of x:

$$x = 1.$$

Substitution of $x = 1$ into the second original equation results in

$$4(1) - y = 2,$$

or

$$4 - y = 2.$$

Isolating y gives

$$-y = 2 - 4,$$

$$-y = -2,$$

and hence,

$$y = 2.$$

The unique solution $x = 1$ and $y = 2$ confirms the validity of the elimination method for this system.

Systems with More Than Two Variables

When addressing systems of equations that extend beyond two variables, the techniques of substitution and elimination continue to be applicable. In a system involving three or more variables, one equation may be manipulated to express one variable in terms of the others. Substituting this expression into the remaining equations effectively reduces the total number of variables. Alternatively, the elimination method may be applied sequentially, canceling one variable at a time until a single-variable equation remains. Through iterative application of these methods, complex systems become manageable, and the values of all unknowns can be determined with systematic precision.

Multiple Choice Questions

1. Which of the following best describes a system of linear equations?

    (a) A set of unrelated equations that can be solved independently.

    (b) A collection of two or more linear equations that share common variables and must be solved simultaneously.

    (c) A single linear equation representing a line on the Cartesian plane.

    (d) A pair of quadratic equations with identical solutions.

2. In the substitution method, what is the primary purpose of isolating one variable in one of the equations?

    (a) To eliminate the need for any further algebraic manipulation.

    (b) To reduce the number of variables in the remaining equation(s) by expressing one variable in terms of another.

    (c) To change the structure of the equation into a quadratic.

    (d) To set up the system for immediate factorization.

3. In the substitution example provided in the chapter, after isolating x in the equation

$$x + 2y = 7,$$

and substituting into
$$3x - y = 5,$$
which equation is obtained before solving for y?

(a) $21 - 7y = 5$

(b) $21 - 6y = 5$

(c) $3(7 - 2y) + y = 5$

(d) $7 - 3y = 5$

4. In the elimination method, why is one of the equations often multiplied by a constant?

   (a) To make the coefficients of one variable equal in magnitude and opposite in sign so that the variable cancels when adding or subtracting.

   (b) To convert the equation into a non-linear form.

   (c) To introduce additional variables for a more robust system.

   (d) To change the equation into a substitution-ready format.

5. In the elimination worked example, the second equation $4x - y = 2$ is multiplied by 3 to yield
$$12x - 3y = 6.$$
When this is added to the first equation $2x + 3y = 8$, what is the resulting equation?

   (a) $14x = 14$

   (b) $14y = 14$

   (c) $2x = 14$

   (d) $12x = 14$

6. When solving a system of equations with three or more variables, which strategy is recommended as discussed in the chapter?

   (a) Use only the substitution method for all equations.

   (b) Apply substitution or elimination iteratively to reduce the system one variable at a time.

(c) Immediately graph the equations to find all intersections.

(d) Use the quadratic formula on every equation.

7. Which of the following is an advantage of the elimination method over the substitution method when solving systems of equations?

(a) It often avoids introducing fractional expressions, making calculations simpler.

(b) It always yields a solution in one step.

(c) It can only be applied to systems with more than two variables.

(d) It eliminates the need to perform any algebraic manipulation.

Answers:

1. **B: A collection of two or more linear equations that share common variables and must be solved simultaneously** This is the definition of a system of linear equations, where each equation involves the same set of variables, and the solution must satisfy all the equations at once.

2. **B: To reduce the number of variables in the remaining equation(s) by expressing one variable in terms of another** In the substitution method, isolating one variable allows you to replace it in the other equation(s), effectively reducing the system to one with fewer variables that is easier to solve.

3. **A: $21 - 7y = 5$** After isolating $x = 7 - 2y$ from the first equation and substituting into $3x - y = 5$, the equation becomes $3(7 - 2y) - y = 5$, which simplifies to $21 - 7y = 5$.

4. **A: To make the coefficients of one variable equal in magnitude and opposite in sign so that the variable cancels when adding or subtracting** Multiplying an equation by a constant adjusts the coefficients so that one of the variables can be eliminated by combining the equations, a key step in the elimination method.

5. **A:** $14x = 14$ Adding the two equations $2x + 3y = 8$ and $12x - 3y = 6$ cancels the y-terms, leading to $14x = 14$, from which x can be easily solved.

6. **B: Apply substitution or elimination iteratively to reduce the system one variable at a time** For systems with more than two variables, the chapter advises using these methods in sequence to gradually reduce the number of variables until the system is simplified enough to solve each variable.

7. **A: It often avoids introducing fractional expressions, making calculations simpler** The elimination method can be advantageous because it frequently allows you to work with whole numbers by canceling out variables directly, whereas substitution may result in fractions when isolating variables.

Practice Problems

1. Solve the following system using the substitution method:

$$x + y = 6,$$

$$2x - y = 1.$$

2. Solve the following system using the elimination method:

$$3x + 4y = 10,$$

$$5x - 4y = 2.$$

3. Solve the following system using the substitution method:

$$2x - 3y = 7,$$
$$x + 2y = 4.$$

4. Solve the following system using the elimination method:

$$4x - y = 9,$$
$$6x + 2y = 14.$$

5. Solve the following system of equations with three variables using the elimination method:

$$x + y + z = 6,$$
$$2x - y + 3z = 14,$$
$$-x + 4y - z = -2.$$

6. Solve the following system using the substitution method:
$$y = 2x + 1,$$
$$3x + 2y = 12.$$

Answers

1. **Problem:** Solve
$$x + y = 6,$$
$$2x - y = 1.$$
Solution: First, solve the first equation for y:
$$y = 6 - x.$$
Substitute this into the second equation:
$$2x - (6 - x) = 1.$$
Simplify the equation:
$$2x - 6 + x = 1 \quad \implies \quad 3x - 6 = 1.$$
Add 6 to both sides:
$$3x = 7,$$
so
$$x = \frac{7}{3}.$$
Now substitute $x = \frac{7}{3}$ back into $y = 6 - x$:
$$y = 6 - \frac{7}{3} = \frac{18}{3} - \frac{7}{3} = \frac{11}{3}.$$

Therefore, the solution is:
$$x = \frac{7}{3}, \quad y = \frac{11}{3}.$$

2. **Problem:** Solve
$$3x + 4y = 10,$$
$$5x - 4y = 2.$$
Solution: Notice that the coefficients of y are opposites. Add the two equations to eliminate y:
$$(3x + 4y) + (5x - 4y) = 10 + 2,$$
which simplifies to:
$$8x = 12.$$
Divide both sides by 8:
$$x = \frac{12}{8} = \frac{3}{2}.$$
Substitute $x = \frac{3}{2}$ into the first equation:
$$3\left(\frac{3}{2}\right) + 4y = 10 \quad \Longrightarrow \quad \frac{9}{2} + 4y = 10.$$
Subtract $\frac{9}{2}$ from both sides:
$$4y = 10 - \frac{9}{2} = \frac{20}{2} - \frac{9}{2} = \frac{11}{2}.$$
Divide by 4:
$$y = \frac{11}{8}.$$
Thus, the solution is:
$$x = \frac{3}{2}, \quad y = \frac{11}{8}.$$

3. **Problem:** Solve
$$2x - 3y = 7,$$
$$x + 2y = 4.$$
Solution: Solve the second equation for x:
$$x = 4 - 2y.$$

Substitute this expression into the first equation:
$$2(4 - 2y) - 3y = 7.$$

Simplify:
$$8 - 4y - 3y = 7 \implies 8 - 7y = 7.$$

Subtract 8 from both sides:
$$-7y = -1,$$

so
$$y = \frac{1}{7}.$$

Now substitute $y = \frac{1}{7}$ back into $x = 4 - 2y$:
$$x = 4 - 2\left(\frac{1}{7}\right) = 4 - \frac{2}{7} = \frac{28}{7} - \frac{2}{7} = \frac{26}{7}.$$

Therefore, the solution is:
$$x = \frac{26}{7}, \quad y = \frac{1}{7}.$$

4. **Problem:** Solve
$$4x - y = 9,$$
$$6x + 2y = 14.$$

Solution: Use the elimination method. First, multiply the first equation by 2 to match the coefficient of y (with opposite sign):
$$2(4x - y) = 2(9) \implies 8x - 2y = 18.$$

Now write the second equation:
$$6x + 2y = 14.$$

Add these two equations:
$$(8x - 2y) + (6x + 2y) = 18 + 14,$$

which simplifies to:
$$14x = 32.$$

Solve for x:
$$x = \frac{32}{14} = \frac{16}{7}.$$

Substitute $x = \frac{16}{7}$ into the first original equation:
$$4\left(\frac{16}{7}\right) - y = 9,$$
$$\frac{64}{7} - y = 9.$$

Solve for y:
$$-y = 9 - \frac{64}{7} = \frac{63}{7} - \frac{64}{7} = -\frac{1}{7},$$
so
$$y = \frac{1}{7}.$$

Thus, the solution is:
$$x = \frac{16}{7}, \quad y = \frac{1}{7}.$$

5. **Problem:** Solve the system
$$x + y + z = 6,$$
$$2x - y + 3z = 14,$$
$$-x + 4y - z = -2.$$

Solution: First, express x from the first equation:
$$x = 6 - y - z.$$

Substitute into the second equation:
$$2(6 - y - z) - y + 3z = 14,$$
$$12 - 2y - 2z - y + 3z = 14,$$
$$12 - 3y + z = 14.$$

Solve for z:
$$z = 14 - 12 + 3y = 2 + 3y.$$

Next, substitute $x = 6 - y - z$ and $z = 2 + 3y$ into the third equation:
$$-(6 - y - (2 + 3y)) + 4y - (2 + 3y) = -2.$$

Simplify inside the parentheses:
$$6 - y - 2 - 3y = 4 - 4y.$$
So the third equation becomes:
$$-(4 - 4y) + 4y - (2 + 3y) = -2.$$
Expand and simplify:
$$-4 + 4y + 4y - 2 - 3y = -2,$$
$$-6 + 5y = -2.$$
Add 6 to both sides:
$$5y = 4 \implies y = \frac{4}{5}.$$
Now, substitute $y = \frac{4}{5}$ into $z = 2 + 3y$:
$$z = 2 + 3\left(\frac{4}{5}\right) = 2 + \frac{12}{5} = \frac{10}{5} + \frac{12}{5} = \frac{22}{5}.$$
Finally, substitute y and z into $x = 6 - y - z$:
$$x = 6 - \frac{4}{5} - \frac{22}{5} = 6 - \frac{26}{5} = \frac{30}{5} - \frac{26}{5} = \frac{4}{5}.$$
Therefore, the solution is:
$$x = \frac{4}{5}, \quad y = \frac{4}{5}, \quad z = \frac{22}{5}.$$

6. **Problem:** Solve
$$y = 2x + 1,$$
$$3x + 2y = 12.$$
Solution: Substitute the expression for y from the first equation into the second equation:
$$3x + 2(2x + 1) = 12.$$
Expand and simplify:
$$3x + 4x + 2 = 12,$$
$$7x + 2 = 12.$$

Subtract 2 from both sides:
$$7x = 10,$$
so
$$x = \frac{10}{7}.$$

Substitute $x = \frac{10}{7}$ back into the first equation:
$$y = 2\left(\frac{10}{7}\right) + 1 = \frac{20}{7} + \frac{7}{7} = \frac{27}{7}.$$

Thus, the solution is:
$$x = \frac{10}{7}, \quad y = \frac{27}{7}.$$

Chapter 20

Applications of Systems of Equations

Model Formulation and Problem Analysis

In many practical scenarios, multiple constraints govern the behavior of a system. When these constraints are expressed through linear relationships, the resulting model is a system of linear equations. Each unknown quantity is represented by a variable, and each constraint is formulated as an equation that captures a relationship between these variables. This approach allows for precise analysis of situations in which several conditions must be simultaneously satisfied.

1 Interpreting Variables and Constraints

When formulating a mathematical model, the first step is to identify the unknown quantities and assign them appropriate variables. A careful analysis of the given constraints determines the coefficients in each equation. For example, in a problem where production processes are limited by machine hours and raw materials, the coefficient multiplying a variable may represent the number of hours required per unit or the amount of material consumed per unit. Such coefficients, derived from the practical context, assist in converting real-world conditions into algebraic expressions that accurately describe the relationships between the variables.

2 Constructing Systems from Real-World Data

After the variables have been defined, the next stage involves translating the conditions of the problem into one or more linear equations. For instance, if the total available resource is known, an equation can be established by summing the contributions (each being the product of a rate and the unknown quantity) of all individual components and equating this sum to the total resource. This process results in a system of equations in which each equation corresponds to a specific constraint. The simultaneous solution of these equations reveals the values of the unknowns that meet all of the imposed conditions.

Worked Example: Production Scheduling in Manufacturing

A manufacturing scenario can serve as an excellent example of how systems of equations are applied to address multiple constraints. In a factory setting, consider the production of two different products, denoted by the variables x and y. Two constraints characterize the production process:

1 Problem Description and System Setup

A limited number of machine hours and a limited supply of raw materials impose restrictions on production. Suppose that each unit of product represented by x requires 2 machine hours, and each unit of product y requires 3 machine hours. If the total available machine hours amount to 35, then the corresponding constraint is expressed as
$$2x + 3y = 35.$$
In addition, assume that production of product x consumes 5 units of raw material per unit, while production of product y consumes 2 units per unit. If the total raw material available is 60 units, the second constraint becomes
$$5x + 2y = 60.$$
Together, these equations form the system that models the production scheduling problem.

2 Solution Process Using the Elimination Technique

To determine the production quantities x and y, the elimination method is applied. First, the first equation is multiplied by 2 in order to facilitate the elimination of one variable:

$$2(2x + 3y) \implies 4x + 6y = 70.$$

Next, the second equation is multiplied by 3:

$$3(5x + 2y) \implies 15x + 6y = 180.$$

Subtracting the modified first equation from the modified second equation eliminates the variable y:

$$(15x + 6y) - (4x + 6y) = 180 - 70.$$

This simplification yields:

$$11x = 110.$$

Dividing both sides by 11 determines the value of x:

$$x = 10.$$

Substituting $x = 10$ into the original equation

$$2x + 3y = 35$$

provides

$$2(10) + 3y = 35 \implies 20 + 3y = 35.$$

Solving for y results in:

$$3y = 15 \implies y = 5.$$

Thus, the production model indicates that 10 units of product x and 5 units of product y satisfy the constraints imposed by available machine hours and raw materials.

Worked Example: Mixture Problem in Chemical Solutions

Systems of equations are equally effective in solving problems that involve combining solutions of varying concentrations. Consider a scenario involving two chemical solutions with different properties.

1 Problem Description and Formulation

Let x represent the volume (in liters) of a solution with a concentration of 10% of a certain substance, and let y represent the volume (in liters) of another solution with a concentration of 30%. The objective is to mix these two solutions to obtain 100 liters of a final solution with a concentration of 20%. The first condition, which imposes the requirement on total volume, is expressed as

$$x + y = 100.$$

The second condition equates the total amount of the substance from the individual solutions to the desired overall amount in the mixture. Given that 10% of the volume x and 30% of the volume y contribute to the substance, and the final mixture must contain 20% of 100 liters (which is 20 liters of the substance), the corresponding equation is

$$0.10x + 0.30y = 20.$$

2 Solution Process Using the Elimination Method

To simplify the process, the second equation is cleared of decimals by multiplying both sides by 10:

$$x + 3y = 200.$$

The resulting system of equations is

$$x + y = 100,$$
$$x + 3y = 200.$$

Subtracting the first equation from the second isolates the variable y:

$$(x + 3y) - (x + y) = 200 - 100,$$
$$2y = 100,$$
$$y = 50.$$

Substituting $y = 50$ back into the equation $x + y = 100$ yields:

$$x = 100 - 50 = 50.$$

This solution indicates that mixing 50 liters of the 10% solution with 50 liters of the 30% solution produces 100 liters of a 20% solution, satisfying both the volume and concentration constraints.

Multiple Choice Questions

1. What is the first essential step when formulating a system of linear equations from a real-world scenario?

    (a) Writing the final solution before modeling
    (b) Defining the unknown variables and identifying the relationships among them
    (c) Graphing the potential equations immediately
    (d) Determining the elimination method to be used

2. In the production scheduling example, the equation $2x + 3y = 35$ represents machine hour constraints. What does the coefficient 2 indicate?

    (a) The total number of machine hours available
    (b) The number of machine hours required per unit of product x
    (c) The number of products x produced per machine hour
    (d) The raw material cost for product x

3. Which description best fits the elimination method used to solve a system of equations?

    (a) Isolating one variable and substituting its value into the other equation
    (b) Multiplying one or both equations by constants so that one variable cancels when the equations are added or subtracted
    (c) Graphing both equations and finding their intersection point
    (d) Rearranging the equations to have identical constant terms

4. When translating real-world data into a system of equations, the numerical coefficients next to the variables most accurately represent:

    (a) The units of measurement for each variable
    (b) The amount of a resource consumed or contributed per unit of the variable

(c) The final solution of the problem

(d) Arbitrarily chosen multipliers for balancing the equation

5. In the mixture problem involving chemical solutions, if x represents the volume of a 10% solution and y represents the volume of a 30% solution, which equation correctly represents the total volume constraint?

   (a) $0.10x + 0.30y = 20$
   (b) $x + y = 100$
   (c) $10x + 30y = 20$
   (d) $x - y = 100$

6. What is the main purpose of multiplying one or both equations by a constant when using the elimination method?

   (a) To simplify the equations into a single variable
   (b) To ensure that the constant terms are equal
   (c) To make the coefficients of one variable identical so that variable can be eliminated
   (d) To convert the system into a graphable form

7. In the production scheduling example, the equation $5x + 2y = 60$ is used to model raw material usage. What does the coefficient 5 most likely represent?

   (a) The total raw materials available for product x
   (b) The number of raw material units needed for product y
   (c) The number of raw material units consumed per unit of product x
   (d) The surplus of raw materials after production

Answers:

1. **B: Defining the unknown variables and identifying the relationships among them**
   Explanation: The very first step in modeling is to determine what the unknown quantities are and how they relate to each other based on the context of the problem.

2. **B: The number of machine hours required per unit of product x**
   Explanation: In the equation $2x + 3y = 35$, the coefficient 2 indicates that each unit of product x consumes 2 machine hours.

3. **B: Multiplying one or both equations by constants so that one variable cancels when the equations are added or subtracted**
   Explanation: The elimination method operates by aligning the coefficients of one of the variables across equations, making it possible to cancel that variable out through addition or subtraction.

4. **B: The amount of a resource consumed or contributed per unit of the variable**
   Explanation: Coefficients in a system of equations are derived from the real-world data and represent measurable quantities, such as rates or resource usage per unit.

5. **B: $x + y = 100$**
   Explanation: The total volume constraint requires that the sum of the volumes of the two solutions equals 100 liters, making $x + y = 100$ the correct equation.

6. **C: To make the coefficients of one variable identical so that variable can be eliminated**
   Explanation: Multiplying by a constant is used so that the coefficients of a selected variable match across the equations. This alignment allows for the elimination of that variable when the equations are combined.

7. **C: The number of raw material units consumed per unit of product x**
   Explanation: In the production scheduling model, the coefficient 5 in the equation $5x + 2y = 60$ represents the number of units of raw material used for each unit of product x, reflecting the real-world constraint on resource consumption.

Practice Problems

1. A gadget factory produces two types of gadgets, X and Y. Each unit of gadget X requires 2 assembly hours and 3 quality

control hours, while each unit of gadget Y requires 3 assembly hours and 2 quality control hours. If the factory has a total of 28 assembly hours and 22 quality control hours available, determine the number of units of each gadget to be produced so that all available hours are used.

Let x = number of gadget X, y = number of gadget Y.

The system is:
$$2x + 3y = 28$$
$$3x + 2y = 22$$

2. A chemist wants to mix two solutions to obtain a 30% acid solution. One solution contains 15% acid and the other contains 45% acid. How many liters of each solution should be mixed to obtain 120 liters of the 30% acid solution?

Let x = liters of 15% solution, y = liters of 45% solution.

Then the equations are:
$$x + y = 120$$
$$0.15x + 0.45y = 36 \quad \text{(since } 0.30 \times 120 = 36\text{)}$$

3. A customer needs to purchase exactly 30 pounds of produce by buying 7 packages. The packages are available in two sizes: one weighing 4 pounds and the other weighing 6 pounds. Determine the number of each type of package the customer should purchase.

$$\text{Let } x = \text{number of 4-pound packages,}$$
$$y = \text{number of 6-pound packages.}$$

The system of equations is:
$$x + y = 7$$
$$4x + 6y = 30$$

4. At a school play, 100 tickets are sold. Adult tickets cost 10 and student tickets cost 6, resulting in a total revenue of 800. Determine the number of adult tickets and student tickets sold.

Let x = number of adult tickets, y = number of student tickets.

The system is:
$$x + y = 100$$
$$10x + 6y = 800$$

5. An investor deposits a total of 10,000 into two funds. Fund A earns 5% annual interest, while Fund B earns 7% annual interest. If the total annual interest earned is 620, how much was invested in each fund?

$$\text{Let } x = \text{amount invested in Fund A,}$$
$$y = \text{amount invested in Fund B.}$$

The system is:
$$x + y = 10000$$
$$0.05x + 0.07y = 620$$

6. A boat travels downstream and upstream. It covers 30 miles downstream in 2 hours and 30 miles upstream in 3 hours. Let v be the boat's speed in still water and c the speed of the current. Formulate a system of equations based on these conditions and determine v and c. Downstream, the effective speed is $v + c$; upstream, it is $v - c$. Thus:

$$v + c = \frac{30}{2} = 15$$
$$v - c = \frac{30}{3} = 10$$

Answers

1. **Solution:** We are given the system:
$$2x + 3y = 28 \quad (1)$$
$$3x + 2y = 22 \quad (2)$$
Multiply equation (1) by 3 and equation (2) by 2 to eliminate x:
$$6x + 9y = 84 \quad (3)$$
$$6x + 4y = 44 \quad (4)$$
Subtract (4) from (3):
$$(6x + 9y) - (6x + 4y) = 84 - 44$$
$$5y = 40 \implies y = 8$$
Substitute $y = 8$ into equation (1):
$$2x + 3(8) = 28 \implies 2x + 24 = 28$$
$$2x = 4 \implies x = 2$$
Explanation: The elimination method helped us cancel out x by aligning coefficients, leading directly to a value for y which was then substituted back to solve for x. Hence, the factory should produce 2 gadgets of type X and 8 gadgets of type Y.

2. **Solution:** The system is:
$$x + y = 120 \quad (1)$$
$$0.15x + 0.45y = 36 \quad (2)$$
To eliminate decimals in (2), multiply the entire equation by 100:
$$15x + 45y = 3600$$
Now, divide by 15:
$$x + 3y = 240 \quad (3)$$
Subtract equation (1) from equation (3):
$$(x + 3y) - (x + y) = 240 - 120$$

222

$$2y = 120 \implies y = 60$$

Substitute $y = 60$ into (1):
$$x + 60 = 120 \implies x = 60$$

Explanation: By converting the percentage equation into an equivalent equation without decimals, we simplified the system. Eliminating x yielded y, and back-substitution provided x. Therefore, 60 liters of the 15% solution and 60 liters of the 45% solution must be mixed.

3. **Solution:** The system is:
$$x + y = 7 \quad (1)$$
$$4x + 6y = 30 \quad (2)$$

From (1), express x in terms of y:
$$x = 7 - y$$

Substitute into (2):
$$4(7 - y) + 6y = 30$$
$$28 - 4y + 6y = 30$$
$$28 + 2y = 30 \implies 2y = 2$$
$$y = 1$$

Now, $x = 7 - 1 = 6$. **Explanation:** By expressing one variable in terms of the other, we reduced the system to a single equation. Solving that equation led to the determination that the customer should purchase 6 packages of 4 pounds and 1 package of 6 pounds.

4. **Solution:** The system is:
$$x + y = 100 \quad (1)$$
$$10x + 6y = 800 \quad (2)$$

Solve (1) for x:
$$x = 100 - y$$

Substitute into (2):
$$10(100 - y) + 6y = 800$$

$$1000 - 10y + 6y = 800$$
$$1000 - 4y = 800 \implies -4y = -200$$
$$y = 50$$

Then, $x = 100 - 50 = 50$. **Explanation:** Substitution allowed us to convert the second equation into one with a single variable. Solving for y and subsequently for x shows that 50 adult tickets and 50 student tickets were sold.

5. **Solution:** The system is:
$$x + y = 10000 \quad (1)$$
$$0.05x + 0.07y = 620 \quad (2)$$

Solve (1) for x:
$$x = 10000 - y$$

Substitute into (2):
$$0.05(10000 - y) + 0.07y = 620$$
$$500 - 0.05y + 0.07y = 620$$
$$500 + 0.02y = 620 \implies 0.02y = 120$$
$$y = \frac{120}{0.02} = 6000$$

Then, $x = 10000 - 6000 = 4000$. **Explanation:** Converting the given interest equation into one variable via substitution and then isolating y led to the conclusion that $6000 was invested in Fund B and $4000 in Fund A.

6. **Solution:** The system is:
$$v + c = 15 \quad (1)$$
$$v - c = 10 \quad (2)$$

Add equations (1) and (2):
$$(v + c) + (v - c) = 15 + 10$$
$$2v = 25 \implies v = 12.5$$

Substitute $v = 12.5$ into (1):
$$12.5 + c = 15 \implies c = 2.5$$

Explanation: Adding the two equations eliminated c directly, giving the boat's speed in still water, v. Substituting back provided the current's speed, c. Thus, the boat's speed in still water is 12.5 mph and the current's speed is 2.5 mph.

Chapter 21

Polynomials: Terminology and Degree

Definition and Fundamental Concepts of Polynomials

A polynomial is an algebraic expression constructed from one or more terms, each of which is a product of a constant and a variable raised to a nonnegative integer exponent. In general, a polynomial in the variable x is expressed in the form

$$a_n x^n + a_{n-1} x^{n-1} + \cdots + a_1 x + a_0,$$

where each a_i is a constant, known as a coefficient, and n is a nonnegative integer representing the highest power of x present in the expression when $a_n \neq 0$. This formal structure provides a clear framework for manipulating and analyzing algebraic expressions.

Terms and Structure in Polynomials

The basic building block of a polynomial is the term, which consists of a coefficient and a power of the variable. A term such as $a_k x^k$ includes the coefficient a_k and the corresponding power k of x. Terms within a polynomial are typically arranged in order of

decreasing exponent, a convention that aids clarity. For example, in the polynomial
$$3x^2 + 5x - 8,$$
the expression comprises three distinct terms: $3x^2$, $5x$, and -8. The organization of terms and the clear delineation of each component form the structural basis for further operations such as addition, subtraction, multiplication, and factorization.

Degree of a Polynomial

The degree of a polynomial is defined as the highest exponent of the variable in the expression, given that the coefficient corresponding to that term is nonzero. In the general form
$$a_n x^n + a_{n-1} x^{n-1} + \cdots + a_1 x + a_0,$$
if $a_n \neq 0$, then the degree of the polynomial is n. For instance, in the polynomial
$$4x^3 + 2x^2 + x + 7,$$
the term $4x^3$ provides the highest exponent and hence the degree of the polynomial is 3. Special cases arise where a constant nonzero polynomial, such as 5, is understood to have degree 0, while the zero polynomial, which contains only the term 0, does not have a defined degree.

Coefficients and Their Roles

The coefficients in a polynomial are the numerical factors that multiply the variable raised to the corresponding power. In an expression of the form
$$a_n x^n + a_{n-1} x^{n-1} + \cdots + a_1 x + a_0,$$
each a_i is a coefficient that quantifies the contribution of the term x^i to the overall expression. Coefficients may be positive, negative, or zero, and they provide crucial information about the magnitude and behavior of the polynomial. For example, in the polynomial
$$-2x^4 + 3x^3 - x + 6,$$
the coefficients are -2, 3, -1, and 6 respectively. The interplay between the coefficients and the associated powers of x determines many of the algebraic properties and the graphical characteristics of the polynomial function.

Multiple Choice Questions

1. What best defines a polynomial?

    (a) An expression that includes variables raised to any power, including negative and fractional.

    (b) An algebraic expression comprised solely of variables without constants.

    (c) An algebraic expression consisting of one or more terms, each being a product of a constant and a variable raised to a nonnegative integer exponent.

    (d) A function that only consists of linear terms.

2. In a polynomial written in standard form, what does the degree represent?

    (a) The coefficient of the highest power term.

    (b) The number of terms in the polynomial.

    (c) The highest exponent of the variable among the terms with nonzero coefficients.

    (d) The constant term at the end of the expression.

3. Which statement is true about a nonzero constant polynomial?

    (a) Its degree is 1.

    (b) Its degree is 0.

    (c) Its degree is the value of the constant.

    (d) Its degree cannot be determined.

4. Consider the polynomial
$$5x^4 - 3x^2 + x - 7.$$
What is its degree?

    (a) 2

    (b) 3

    (c) 4

    (d) 5

5. In the polynomial
$$-2x^3 + 0x^2 + 4,$$
what is the role of the term $0x^2$ in determining the degree?

   (a) It lowers the degree to 2.
   (b) It is counted, making the degree 2 since it appears in the expression.
   (c) It is ignored because its coefficient is zero, so it does not affect the degree.
   (d) It is considered equivalent to a constant term.

6. What does the term "coefficient" mean in the context of polynomial terms?

   (a) The variable part of a term.
   (b) The exponent indicating how many times the variable is used.
   (c) The numerical factor multiplied by the variable raised to an exponent.
   (d) The sum of all the variable parts in a polynomial.

7. Why are polynomials typically written with terms in descending order of exponents?

   (a) It simplifies the process of adding and subtracting polynomials.
   (b) It emphasizes the highest degree term, which determines the polynomial's degree and overall behavior.
   (c) It makes the polynomial easier to factor.
   (d) It ensures the constant term is displayed last.

Answers:

1. **C: An algebraic expression consisting of one or more terms, each being a product of a constant and a variable raised to a nonnegative integer exponent.**
This is the formal definition of a polynomial, distinguishing it from expressions that include negative, fractional, or variable-only components.

2. **C: The highest exponent of the variable among the terms with nonzero coefficients.**
   The degree is defined by the term with the largest exponent (and nonzero coefficient), which determines many of the polynomial's properties.

3. **B: Its degree is 0.**
   A nonzero constant can be seen as a polynomial with one term, where the variable is raised to the 0 power (since $x^0 = 1$), making its degree 0.

4. **C: 4.**
   In the polynomial $5x^4 - 3x^2 + x - 7$, the term $5x^4$ has the highest exponent (4), so the degree is 4.

5. **C: It is ignored because its coefficient is zero, so it does not affect the degree.**
   Terms with a zero coefficient do not contribute to the value of the polynomial and are not considered when determining its degree.

6. **C: The numerical factor multiplied by the variable raised to an exponent.**
   In any term of the form $a_k x^k$, a_k is the coefficient, indicating how much the variable's term contributes to the overall expression.

7. **B: It emphasizes the highest degree term, which determines the polynomial's degree and overall behavior.**
   Writing polynomials in descending order helps to quickly identify the highest power, clarifying the degree and simplifying further operations like addition, subtraction, and factoring.

Practice Problems

1. Define a polynomial. In your answer, include the roles of terms, coefficients, and exponents, and provide a simple example.

2. Determine the degree of the following polynomial:
$$4x^3 + 2x^2 - 8x + 5.$$

3. Given the constant polynomial
$$7,$$
explain what its degree is and justify your answer.

4. For the polynomial
$$-3x^4 + 5x^2 - x + 9,$$
list each term, identify the coefficient of each term, and state the degree of the polynomial.

5. Explain why it is customary to arrange the terms of a polynomial in order of decreasing exponents. How does this arrangement facilitate operations with polynomials?

6. Discuss the definition of the zero polynomial and explain why its degree is considered undefined.

Answers

1. **Solution:** A polynomial is an algebraic expression constructed from one or more terms, where each term is the product of a constant (called a coefficient) and a variable raised to a nonnegative integer exponent. The principal components of a polynomial are:

    - **Terms:** These are the individual parts of the expression (for example, in ax^n, the entire expression is one term).
    - **Coefficients:** These are the numerical factors multiplying the variable parts.
    - **Exponents:** These indicate the power to which the variable is raised, and must be nonnegative integers.

For example, consider the polynomial
$$3x^2 - 2x + 7.$$
It consists of three terms: $3x^2$ (with coefficient 3 and exponent 2), $-2x$ (with coefficient -2 and exponent 1), and 7 (which can be seen as $7x^0$ with exponent 0).

2. **Solution:** The degree of a polynomial is defined as the highest exponent of the variable in any term with a nonzero coefficient. In the polynomial
$$4x^3 + 2x^2 - 8x + 5,$$
the term $4x^3$ has the highest exponent, which is 3. Therefore, the degree of the polynomial is 3.

3. **Solution:** The polynomial
$$7$$
is a constant polynomial. A constant can be expressed as $7x^0$, where $x^0 = 1$. Since the only term has an exponent of 0, the degree of a nonzero constant polynomial is 0.

4. **Solution:** Consider the polynomial
$$-3x^4 + 5x^2 - x + 9.$$
We can break it down as follows:
- Term 1: $-3x^4$ with coefficient -3 and exponent 4.
- Term 2: $5x^2$ with coefficient 5 and exponent 2.
- Term 3: $-x$ which is equivalent to $-1x^1$ with coefficient -1 and exponent 1.
- Term 4: 9 which can be considered as $9x^0$ with coefficient 9 and exponent 0.

The highest exponent among these terms is 4 (from $-3x^4$), so the degree of the polynomial is 4.

5. **Solution:** Arranging the terms of a polynomial in order of decreasing exponents is customary because:
- It allows for immediate identification of the degree of the polynomial, as the first term will have the highest exponent.

- It facilitates the process of adding, subtracting, or comparing polynomials, since like terms (terms with the same exponent) can be easily aligned.
- It simplifies operations such as polynomial long division or synthetic division by providing a systematic format.

6. **Solution:** The zero polynomial is defined as the polynomial in which all coefficients are zero, typically denoted by

$$0.$$

For any nonzero polynomial, the degree is the highest exponent with a nonzero coefficient. However, since the zero polynomial does not contain any nonzero terms, it does not have a well-defined degree. In many contexts, the degree of the zero polynomial is considered undefined (or sometimes defined as negative infinity) because it does not fit within the standard framework used for other polynomials.

Chapter 22

Operations with Polynomials

Addition of Polynomials

Polynomial addition is performed by combining terms that share identical variable components and exponents. Each term within a polynomial is expressed as a product of a coefficient and a variable raised to a nonnegative integer exponent. To add two or more polynomials, it is necessary to align the terms having the same power of the variable and then sum their coefficients. This process of combining like terms ensures that the resulting polynomial is organized and simplified.

For example, consider the polynomials

$$P(x) = 3x^2 + 2x - 5 \quad \text{and} \quad Q(x) = 5x^2 - x + 4.$$

To add these expressions, begin by grouping the like terms:

$$(3x^2 + 5x^2) + (2x - x) + (-5 + 4).$$

The sum of the coefficients for the x^2 terms is $3 + 5 = 8$, for the x terms it is $2 - 1 = 1$, and for the constant terms it is $-5 + 4 = -1$. Hence, the resulting polynomial from the addition is

$$P(x) + Q(x) = 8x^2 + x - 1.$$

Subtraction of Polynomials

Subtraction of polynomials is closely related to addition and requires careful management of the sign change when subtracting an entire polynomial. The procedure involves distributing the negative sign across the polynomial being subtracted, followed by combining the like terms that result.

Consider the polynomials
$$A(x) = 6x^3 + 4x - 2 \quad \text{and} \quad B(x) = 3x^3 - 2x + 1.$$

Express the subtraction as
$$A(x) - B(x) = 6x^3 + 4x - 2 - (3x^3 - 2x + 1).$$

Distributing the negative sign yields
$$6x^3 + 4x - 2 - 3x^3 + 2x - 1.$$

Next, combine like terms by grouping the coefficients of the corresponding terms:
$$(6x^3 - 3x^3) + (4x + 2x) + (-2 - 1).$$

This simplifies to
$$3x^3 + 6x - 3.$$

Thus, the simplified expression for the subtraction is
$$A(x) - B(x) = 3x^3 + 6x - 3.$$

Multiplication of Polynomials

Multiplication of polynomials relies on the distributive property, which necessitates that every term in the first polynomial is multiplied by each term in the second polynomial. This process involves the multiplication of coefficients and the addition of exponents when variables of the same base are multiplied together. After obtaining all the individual products, the resulting terms are combined by grouping like terms together, thereby achieving a simplified expression.

Take, for instance, the multiplication of the polynomials
$$M(x) = 2x + 3 \quad \text{and} \quad N(x) = x^2 - x + 4.$$

The distributive property is applied as follows:

$(2x) \cdot (x^2) + (2x) \cdot (-x) + (2x) \cdot 4 + (3) \cdot (x^2) + (3) \cdot (-x) + (3) \cdot 4.$

Calculating each product separately gives

$$2x^3 - 2x^2 + 8x + 3x^2 - 3x + 12.$$

The next step involves combining like terms. Note that the single x^3 term remains as is, while the x^2 terms $-2x^2$ and $3x^2$ combine to yield x^2, and the x terms $8x$ and $-3x$ combine to yield $5x$. The constant term is 12. Therefore, the product of the polynomials simplifies to

$$M(x) \cdot N(x) = 2x^3 + x^2 + 5x + 12.$$

By consistently aligning and organizing like terms during each operation, the clarity and simplicity of the resulting polynomial expressions are maintained throughout the processes of addition, subtraction, and multiplication.

Multiple Choice Questions

1. When adding two polynomials, what is the most important step to obtain the correct result?

    (a) Multiplying the coefficients of the terms.

    (b) Combining like terms by adding their coefficients.

    (c) Distributing the variable across all terms.

    (d) Subtracting the exponents of similar terms.

2. When subtracting one polynomial from another, what is the first necessary step?

    (a) Reordering the terms of the subtracted polynomial.

    (b) Multiplying the entire subtracted polynomial by -1.

    (c) Adding the corresponding coefficients without change.

    (d) Dividing each term by the highest degree term.

3. Which algebraic property is primarily used when multiplying two polynomials?

    (a) Commutative property.

(b) Associative property.

  (c) Distributive property.

  (d) Identity property.

4. Which of the following best defines like terms in a polynomial expression?

    (a) Terms with the same numerical coefficient.

    (b) Terms that appear next to each other.

    (c) Terms having identical variable parts with the same exponents.

    (d) Terms with the same sign.

5. Determine the sum of the polynomials:

    $P(x) = 2x^2 + 3x - 4$ and $Q(x) = 5x^2 - 2x + 6$.

    (a) $7x^2 + x + 2$

    (b) $7x^2 + 5x + 2$

    (c) $7x^2 + x - 10$

    (d) $7x^2 - x + 10$

6. Compute the simplified expression for the subtraction:

    $A(x) - B(x) = (6x^3 + 4x - 2) - (3x^3 - 2x + 1)$.

    (a) $3x^3 + 2x - 1$

    (b) $3x^3 + 6x - 3$

    (c) $9x^3 + 2x - 1$

    (d) $3x^3 + 2x + 3$

7. What is the result of multiplying the polynomials

    $M(x) = x + 2$ and $N(x) = x^2 - 3x + 4$?

    (a) $x^3 - x^2 - 2x + 8$

    (b) $x^3 - x^2 + 2x + 8$

    (c) $x^3 + x^2 - 2x + 8$

    (d) $x^3 + x^2 + 2x + 8$

8. Which of the following steps is common to both addition and subtraction of polynomials?

(a) Arranging the terms in descending order of degree.

(b) Multiplying the coefficients of like terms.

(c) Combining like terms.

(d) Distributing any constants across the polynomial.

Answers:

1. **B: Combining like terms by adding their coefficients**
   When adding polynomials, you must identify and combine terms that have the same variables raised to the same powers in order to simplify the expression.

2. **B: Multiplying the entire subtracted polynomial by -1**
   Distributing the negative sign across all terms in the polynomial being subtracted is crucial; it changes the sign of each term before combining like terms.

3. **C: Distributive property**
   Multiplying polynomials requires that you multiply each term in the first polynomial by every term in the second polynomial, a process that is fundamentally based on the distributive property.

4. **C: Terms having identical variable parts with the same exponents**
   Like terms are defined as those that have exactly the same variable factors raised to the same powers; only their coefficients may differ.

5. **A:** $7x^2 + x + 2$
   Adding the like terms: $2x^2 + 5x^2 = 7x^2$, $3x - 2x = x$, and $-4 + 6 = 2$; hence, the sum is $7x^2 + x + 2$.

6. **B:** $3x^3 + 6x - 3$
   Subtracting the polynomials involves distributing the negative sign: $6x^3 - 3x^3 = 3x^3$, $4x + 2x = 6x$, and $-2 - 1 = -3$, leading to $3x^3 + 6x - 3$.

7. **A:** $x^3 - x^2 - 2x + 8$
   Multiplying $(x + 2)$ by $(x^2 - 3x + 4)$ gives $x \cdot (x^2 - 3x + 4) = x^3 - 3x^2 + 4x$ and $2 \cdot (x^2 - 3x + 4) = 2x^2 - 6x + 8$; combining like terms yields $x^3 - x^2 - 2x + 8$.

8. **C: Combining like terms**
   In both addition and subtraction of polynomials, after aligning the terms (or after distributing a negative sign in subtraction), the next essential step is to combine like terms by adding their coefficients.

Practice Problems

1. Add the following polynomials:

   $P(x) = 4x^3 - 2x^2 + x - 7$ and $Q(x) = 3x^3 + 5x^2 - 4x + 2$.

2. Subtract the polynomial

   $$Q(x) = 2x^4 - 4x^2 + 7$$

   from the polynomial

   $$R(x) = 5x^4 + 3x^2 - x + 8.$$

3. Multiply the following polynomials:
$$T(x) = x + 2 \quad \text{and} \quad U(x) = 2x^2 - 3x + 4.$$

4. Compute the following expression by first adding the polynomials
$$A(x) = x^2 + 2x \quad \text{and} \quad B(x) = 3x^2 - x + 5,$$
then multiplying the result by the polynomial
$$C(x) = 2x + 4.$$

5. Simplify the expression by performing the operations in the correct order:
$$E(x) = (2x + 3) - (x^2 - 4x + 1) + (3x^2 - 2x - 5).$$

6. **Multiply the quadratic polynomials:**

$$F(x) = x^2 - 3x + 2 \quad \text{and} \quad G(x) = x^2 + 5x - 6.$$

Answers

1. **Addition of Polynomials:**

   To add $P(x)$ and $Q(x)$, we combine like terms (terms with the same power of x). Write the sum as

   $$P(x)+Q(x) = (4x^3+3x^3)+(-2x^2+5x^2)+(x-4x)+(-7+2).$$

   Simplify each set of like terms:

   $$4x^3 + 3x^3 = 7x^3, \quad -2x^2 + 5x^2 = 3x^2,$$

   $$x - 4x = -3x, \quad -7 + 2 = -5.$$

   Thus, the resulting polynomial is

   $$P(x) + Q(x) = 7x^3 + 3x^2 - 3x - 5.$$

2. **Subtraction of Polynomials:**

   To find $R(x) - Q(x)$, write the subtraction as

   $$R(x) - Q(x) = \left(5x^4 + 3x^2 - x + 8\right) - \left(2x^4 - 4x^2 + 7\right).$$

   Distribute the negative sign:

   $$= 5x^4 + 3x^2 - x + 8 - 2x^4 + 4x^2 - 7.$$

241

Now, combine like terms:
$$5x^4 - 2x^4 = 3x^4, \quad 3x^2 + 4x^2 = 7x^2, \quad -x \text{ remains}, \quad 8-7 = 1.$$

Therefore, the simplified result is
$$R(x) - Q(x) = 3x^4 + 7x^2 - x + 1.$$

3. **Multiplication of Polynomials:**

Multiplying $T(x)$ by $U(x)$ requires multiplying every term in $T(x)$ by every term in $U(x)$:
$$(x+2)(2x^2 - 3x + 4) = x(2x^2 - 3x + 4) + 2(2x^2 - 3x + 4).$$

Compute the products:
$$x \cdot 2x^2 = 2x^3, \quad x \cdot (-3x) = -3x^2, \quad x \cdot 4 = 4x,$$
$$2 \cdot 2x^2 = 4x^2, \quad 2 \cdot (-3x) = -6x, \quad 2 \cdot 4 = 8.$$

Combine like terms:
$$2x^3, \quad (-3x^2 + 4x^2) = x^2, \quad (4x - 6x) = -2x, \quad \text{constant } 8.$$

The final simplified product is
$$T(x) \cdot U(x) = 2x^3 + x^2 - 2x + 8.$$

4. **Combined Operations:**

First, add the polynomials $A(x)$ and $B(x)$:
$$A(x) + B(x) = (x^2 + 2x) + (3x^2 - x + 5) = (x^2 + 3x^2) + (2x - x) + 5.$$

Simplify the sum:
$$A(x) + B(x) = 4x^2 + x + 5.$$

Next, multiply this result by $C(x)$:
$$(4x^2 + x + 5)(2x + 4).$$

Distribute each term:
$$4x^2 \cdot 2x = 8x^3, \quad 4x^2 \cdot 4 = 16x^2,$$
$$x \cdot 2x = 2x^2, \quad x \cdot 4 = 4x,$$

$$5 \cdot 2x = 10x, \quad 5 \cdot 4 = 20.$$

Combine like terms:

$$8x^3, \quad 16x^2 + 2x^2 = 18x^2, \quad 4x + 10x = 14x, \quad \text{and } 20.$$

Therefore, the overall expression becomes

$$(A(x) + B(x)) \cdot C(x) = 8x^3 + 18x^2 + 14x + 20.$$

5. **Simplification Involving Addition and Subtraction:**
   Begin with the expression

   $$E(x) = (2x + 3) - (x^2 - 4x + 1) + (3x^2 - 2x - 5).$$

   First, distribute the negative sign:

   $$= 2x + 3 - x^2 + 4x - 1 + 3x^2 - 2x - 5.$$

   Group and combine like terms:

   $$\text{Quadratic terms: } -x^2 + 3x^2 = 2x^2,$$

   $$\text{Linear terms: } 2x + 4x - 2x = 4x,$$

   $$\text{Constants: } 3 - 1 - 5 = -3.$$

   The simplified expression is

   $$E(x) = 2x^2 + 4x - 3.$$

6. **Multiplication of Quadratic Polynomials:**
   To multiply $F(x)$ and $G(x)$, use the distributive property:

   $$(x^2 - 3x + 2)(x^2 + 5x - 6).$$

   Step 1: Multiply the terms in the first polynomial by x^2:

   $$x^2 \cdot x^2 = x^4, \quad x^2 \cdot 5x = 5x^3, \quad x^2 \cdot (-6) = -6x^2.$$

   Step 2: Multiply the terms by $-3x$:

   $$-3x \cdot x^2 = -3x^3, \quad -3x \cdot 5x = -15x^2, \quad -3x \cdot (-6) = 18x.$$

   Step 3: Multiply the terms by 2:

   $$2 \cdot x^2 = 2x^2, \quad 2 \cdot 5x = 10x, \quad 2 \cdot (-6) = -12.$$

Now, combine like terms:
$$x^4,$$
Cubic terms: $5x^3 - 3x^3 = 2x^3$,

Quadratic terms: $-6x^2 - 15x^2 + 2x^2 = -19x^2$,

Linear terms: $18x + 10x = 28x$,

Constant term: -12.

Hence, the final product is
$$F(x) \cdot G(x) = x^4 + 2x^3 - 19x^2 + 28x - 12.$$

Chapter 23

Special Products of Polynomials

Square of a Binomial

The square of a binomial is obtained by multiplying a binomial by itself. For an expression of the form

$$(a + b),$$

the square is given by

$$(a + b)^2 = a^2 + 2ab + b^2.$$

In this expansion, the first term is the square of the first component, the last term is the square of the second component, and the middle term is twice the product of the two components. When the binomial has a subtraction sign, the square is

$$(a - b)^2 = a^2 - 2ab + b^2.$$

For example, the expansion of

$$(x + 3)^2$$

is computed as

$$(x + 3)^2 = x^2 + 2 \cdot x \cdot 3 + 3^2 = x^2 + 6x + 9.$$

This perfect square trinomial form greatly simplifies further manipulation of polynomial expressions.

Product of a Sum and a Difference

Multiplying the sum and difference of the same two terms leads to a remarkable cancellation of the middle terms. Considering the expression
$$(a + b)(a - b),$$
the expansion is given by
$$(a + b)(a - b) = a^2 - b^2.$$
The process involves distributing each term in the first binomial across the second binomial. The intermediate products, ab and $-ab$, cancel each other, leaving the difference of the squares of the original terms. An illustrative example is the product
$$(2x + 5)(2x - 5),$$
which simplifies as follows:
$$(2x + 5)(2x - 5) = (2x)^2 - 5^2 = 4x^2 - 25.$$
This identity is particularly useful in simplifying expressions and solving equations where the difference of squares pattern appears.

Cube of a Binomial

Expanding the cube of a binomial involves a systematic combination of powers and coefficients. The cube of a sum is expressed by the formula
$$(a + b)^3 = a^3 + 3a^2b + 3ab^2 + b^3,$$
and correspondingly, the cube of a difference is given by
$$(a - b)^3 = a^3 - 3a^2b + 3ab^2 - b^3.$$
The coefficients in these expansions are derived from the binomial coefficients present in Pascal's Triangle. An example is provided by the expansion of
$$(x - 2)^3,$$
which computes as
$$(x - 2)^3 = x^3 - 3x^2 \cdot 2 + 3x \cdot 2^2 - 2^3 = x^3 - 6x^2 + 12x - 8.$$
This cubic expansion is an important tool for recognizing and simplifying expressions that involve raised binomials.

Sum and Difference of Cubes

Expressions involving the sum or difference of cubes can be factored using specific formulas that decompose the cubic expression into a product consisting of a binomial and a quadratic polynomial. The sum of cubes is factored as

$$a^3 + b^3 = (a+b)(a^2 - ab + b^2),$$

and the difference of cubes is factored as

$$a^3 - b^3 = (a-b)(a^2 + ab + b^2).$$

These identities are particularly useful when reducing cubic equations to forms that are simpler to manipulate or solve. For instance, the expression

$$8x^3 + 27$$

can be written as

$$(2x)^3 + 3^3,$$

and then factored according to the sum of cubes formula:

$$8x^3 + 27 = (2x+3)\left((2x)^2 - 2x \cdot 3 + 3^2\right) = (2x+3)(4x^2 - 6x + 9).$$

Similarly, a difference of cubes such as

$$x^3 - 64,$$

where

$$64 = 4^3,$$

is factored as

$$x^3 - 64 = (x-4)(x^2 + 4x + 16).$$

These factorization strategies alleviate the complexity of handling cubic expressions by revealing their underlying structure, thereby simplifying subsequent algebraic operations.

Multiple Choice Questions

1. Which of the following is the correct expansion of the square of a binomial (a + b)²?

(a) $a^2 + ab + b^2$
(b) $a^2 + 2ab + b^2$
(c) $a^2 + b^2$
(d) $a^2 - 2ab + b^2$

2. What is the proper expansion for the square of the binomial $(a - b)^2$?

   (a) $a^2 + 2ab + b^2$
   (b) $a^2 - 2ab + b^2$
   (c) $a^2 + b^2 - 2ab$
   (d) $a^2 - b^2$

3. The product of the sum and difference, $(a + b)(a - b)$, simplifies to:

   (a) $a^2 + b^2$
   (b) $a^2 - b^2$
   (c) $a^2 + 2ab + b^2$
   (d) $a^2 - 2ab + b^2$

4. Which of the following represents the correct expansion of the cube of a binomial $(a - b)^3$?

   (a) $a^3 - 3a^2b + 3ab^2 - b^3$
   (b) $a^3 + 3a^2b + 3ab^2 + b^3$
   (c) $a^3 - 3a^2b - 3ab^2 + b^3$
   (d) $a^3 + 3a^2b - 3ab^2 - b^3$

5. Which identity correctly factors a sum of cubes, $a^3 + b^3$?

   (a) $a^3 + b^3 = (a + b)(a^2 + ab + b^2)$
   (b) $a^3 + b^3 = (a + b)(a^2 - ab + b^2)$
   (c) $a^3 + b^3 = (a - b)(a^2 + ab + b^2)$
   (d) $a^3 + b^3 = (a - b)(a^2 - ab + b^2)$

6. The expression $8x^3 + 27$ is a sum of cubes. Which factorization is correct?

   (a) $(2x + 3)(4x^2 + 6x + 9)$
   (b) $(2x + 3)(4x^2 - 6x + 9)$

(c) $(4x + 3)(2x^2 - 6x + 9)$
(d) $(2x - 3)(4x^2 - 6x + 9)$

7. Which factorization identity correctly applies to the difference of cubes $a^3 - b^3$?

(a) $a^3 - b^3 = (a + b)(a^2 - ab + b^2)$
(b) $a^3 - b^3 = (a - b)(a^2 - ab + b^2)$
(c) $a^3 - b^3 = (a - b)(a^2 + ab + b^2)$
(d) $a^3 - b^3 = (a + b)(a^2 + ab + b^2)$

Answers:

1. **B: $a^2 + 2ab + b^2$**
   Explanation: The expansion follows the pattern $(a + b)^2 = a^2 + 2ab + b^2$, where the middle term arises from adding two identical products of ab.

2. **B: $a^2 - 2ab + b^2$**
   Explanation: In $(a - b)^2$, the negative sign affects the middle term, resulting in -2ab, while both squared terms remain positive.

3. **B: $a^2 - b^2$**
   Explanation: Multiplying $(a + b)(a - b)$ gives $a^2 - b^2$ because the middle terms, ab and -ab, cancel each other.

4. **A: $a^3 - 3a^2b + 3ab^2 - b^3$**
   Explanation: The expansion of $(a - b)^3$ follows from the binomial theorem with alternating signs: a^3 first, then $-3a^2b$, followed by $+3ab^2$, and finally $-b^3$.

5. **B: $a^3 + b^3 = (a + b)(a^2 - ab + b^2)$**
   Explanation: The sum of cubes factors with the identity $a^3 + b^3 = (a + b)(a^2 - ab + b^2)$, which breaks the cubic expression into a linear and a quadratic factor.

6. **B: $(2x + 3)(4x^2 - 6x + 9)$**
   Explanation: Recognizing $8x^3$ as $(2x)^3$ and 27 as 3^3, the sum of cubes factorization yields $(2x + 3)((2x)^2 - (2x)(3) + 3^2)$ which simplifies to $(2x + 3)(4x^2 - 6x + 9)$.

7. **C: $a^3 - b^3 = (a - b)(a^2 + ab + b^2)$**
   Explanation: The difference of cubes is factored using the identity $a^3 - b^3 = (a - b)(a^2 + ab + b^2)$, which ensures the correct distribution of terms for proper factorization.

Practice Problems

1. Expand the square of the binomial:
$$(x+3)^2$$

2. Expand the square of the binomial with subtraction:
$$(2x-5)^2$$

3. Expand the product of a sum and a difference:
$$(4x+7)(4x-7)$$

4. Expand the cube of a binomial:
$$(x-2)^3$$

5. Factor the sum of cubes:
$$27x^3 + 8$$

6. Factor the difference of cubes:
$$x^3 - 64$$

Answers

1. Expand the square of the binomial:
$$(x+3)^2$$

   Solution:
   The formula for the square of a binomial is
   $$(a+b)^2 = a^2 + 2ab + b^2.$$
   Here, $a = x$ and $b = 3$. Substituting, we have:
   $$(x+3)^2 = x^2 + 2 \cdot x \cdot 3 + 3^2 = x^2 + 6x + 9.$$
   Therefore,
   $$(x+3)^2 = x^2 + 6x + 9.$$

2. Expand the square of the binomial with subtraction:
$$(2x-5)^2$$

   Solution:
   The formula for the square of a binomial with a subtraction is
   $$(a-b)^2 = a^2 - 2ab + b^2.$$
   Here, $a = 2x$ and $b = 5$. Thus,
   $$(2x-5)^2 = (2x)^2 - 2 \cdot (2x) \cdot 5 + 5^2.$$
   Computing each term, we find:
   $$(2x)^2 = 4x^2, \quad 2 \cdot (2x) \cdot 5 = 20x, \quad 5^2 = 25.$$
   Hence,
   $$(2x-5)^2 = 4x^2 - 20x + 25.$$

3. Expand the product of a sum and a difference:
$$(4x+7)(4x-7)$$

   Solution:
   This product follows the difference of squares formula:
   $$(a+b)(a-b) = a^2 - b^2.$$

Let $a = 4x$ and $b = 7$. Then,
$$(4x+7)(4x-7) = (4x)^2 - 7^2 = 16x^2 - 49.$$

Therefore,
$$(4x+7)(4x-7) = 16x^2 - 49.$$

4. Expand the cube of a binomial:
$$(x-2)^3$$

Solution:
The cube of a binomial is expanded using the formula:
$$(a-b)^3 = a^3 - 3a^2b + 3ab^2 - b^3.$$

Here, $a = x$ and $b = 2$. Substituting into the formula, we get:
$$(x-2)^3 = x^3 - 3x^2(2) + 3x(2)^2 - (2)^3.$$

Simplify each term:
$$-3x^2(2) = -6x^2, \quad 3x(2)^2 = 3x \cdot 4 = 12x, \quad (2)^3 = 8.$$

Thus,
$$(x-2)^3 = x^3 - 6x^2 + 12x - 8.$$

5. Factor the sum of cubes:
$$27x^3 + 8$$

Solution:
Recognize that $27x^3 = (3x)^3$ and $8 = 2^3$. The sum of cubes can be factored using:
$$a^3 + b^3 = (a+b)(a^2 - ab + b^2).$$

Here, $a = 3x$ and $b = 2$. Substituting, we have:
$$27x^3 + 8 = (3x+2)\left[(3x)^2 - (3x)(2) + 2^2\right].$$

Simplify the quadratic factor:
$$(3x)^2 = 9x^2, \quad (3x)(2) = 6x, \quad 2^2 = 4.$$

Therefore,
$$27x^3 + 8 = (3x+2)(9x^2 - 6x + 4).$$

6. Factor the difference of cubes:
$$x^3 - 64$$

Solution:
Notice that $64 = 4^3$. The difference of cubes formula is:
$$a^3 - b^3 = (a-b)(a^2 + ab + b^2).$$

Here, $a = x$ and $b = 4$. Thus,
$$x^3 - 64 = (x-4)\left[x^2 + 4x + 4^2\right].$$

Since $4^2 = 16$, we get:
$$x^3 - 64 = (x-4)(x^2 + 4x + 16).$$

Chapter 24

Factoring Techniques for Higher-Degree Polynomials

Factoring by Grouping

In many instances, a polynomial with several terms does not allow for factoring by a common factor across every term but can be rearranged into groups that share common factors. This method, known as factoring by grouping, involves partitioning the polynomial into two or more sets of terms. Each group is then factored separately and the resulting expression is examined for a common binomial factor.

Consider a polynomial of four terms written as

$$ax^3 + bx^2 + cx + d.$$

The first step is to identify a natural grouping. Frequently, the polynomial can be divided into two pairs:

$$(ax^3 + bx^2) + (cx + d).$$

Within the first grouping, a common factor (often a power of x and a numerical coefficient) is factored out, yielding an expression of the form

$$x^2(a'x + b'),$$

where a' and b' denote the resulting coefficients after factoring. Similarly, in the second grouping, a constant factor is extracted so that the terms become
$$c'(x + d'),$$
with the understanding that the binomial $(x + \text{constant})$ matches that found in the first group. When such a factor is common between the groups, the polynomial factors further as
$$(x + \text{constant})\Big(x^2 \cdot (\text{coefficient}) + (\text{other coefficient})\Big).$$

A typical example that illustrates this method is the polynomial
$$x^3 + 2x^2 - x - 2.$$
Grouping the terms as
$$(x^3 + 2x^2) + (-x - 2),$$
the first group factors as
$$x^2(x + 2),$$
while the second group factors as
$$-1(x + 2).$$
The common factor $(x + 2)$ is then factored out:
$$(x + 2)(x^2 - 1).$$
The expression $x^2 - 1$ is itself a difference of squares and factors further, but the grouping method is demonstrated by the process of extracting a common binomial factor.

Polynomials with six or more terms may require grouping into three or more sets. In such cases, a careful analysis of the coefficients and term structure is essential in order to identify groupings that yield a common binomial factor. The order of the terms can sometimes be adjusted without changing the overall expression in order to enhance the visibility of common factors.

Pattern Recognition in Polynomial Factoring

Higher-degree polynomials often present themselves in forms that, upon closer inspection, resemble patterns known from lower-degree cases. Recognizing these patterns is a key strategy in simplifying and factoring complex expressions.

One common pattern arises when the exponents of x in a polynomial follow a regular sequence, thereby allowing the expression to be treated as a quadratic in another variable. For example, an expression such as
$$x^4 + 5x^2 + 6$$
may be transformed by setting $u = x^2$, converting it to a quadratic equation:
$$u^2 + 5u + 6.$$
Once factored in terms of u as
$$(u+2)(u+3),$$
the original variable is reintroduced by substituting x^2 back for u:
$$(x^2 + 2)(x^2 + 3).$$
This substitution technique is particularly effective when the exponents of the polynomial are multiples of a lower exponent, allowing the expression to mirror the standard quadratic form.

In addition, many higher-degree polynomials exhibit features such as differences of squares or sums and differences of cubes. Recognition of these familiar forms provides an immediate pathway to factorization. An expression taking the form
$$a^4 - b^4$$
may be recognized as a difference of squares because it can be written as
$$(a^2)^2 - (b^2)^2.$$
Applying the difference of squares identity yields:
$$(a^2 - b^2)(a^2 + b^2),$$
and further factoring of $a^2 - b^2$ leads to
$$(a-b)(a+b)(a^2 + b^2).$$

Another common pattern appears in polynomials that are nearly symmetric in their terms. Such expressions may be rearranged to reveal hidden quadratic-like structures or other well-known forms. The careful observation of coefficients and exponents is central to this technique, as it requires discerning whether a polynomial fits into a recognizable pattern such as a perfect square trinomial or a cubic identity.

Factoring by Substitution

Substitution as a method enables the reduction of a higher-degree polynomial into a more familiar form. When exponents of the variable appear in a repeating manner, substitution reduces complexity and assists in the identification of factors.

For instance, consider the polynomial
$$x^6 + 7x^3 + 12.$$

A substitution is made by letting
$$u = x^3.$$

This change transforms the expression into a quadratic in u:
$$u^2 + 7u + 12.$$

The quadratic factors in the conventional manner as
$$(u+3)(u+4).$$

Upon reverting to the original variable by replacing u with x^3, the factorization becomes
$$(x^3 + 3)(x^3 + 4).$$

This method proves valuable when exponents follow a common multiple, enabling a seamless transition from a difficult-to-factor polynomial to one that conforms to a standard and well-understood form.

Another example involves a biquadratic polynomial such as
$$x^4 - 5x^2 + 4.$$

Setting
$$u = x^2,$$
recasts the polynomial into
$$u^2 - 5u + 4,$$
which factors as
$$(u-1)(u-4).$$

Substituting back yields the factorization
$$(x^2 - 1)(x^2 - 4),$$
and further recognizing the differences of squares gives:
$$(x - 1)(x + 1)(x - 2)(x + 2).$$

The substitution technique not only simplifies the process but also facilitates the application of familiar quadratic factoring methods. It is particularly effective when polynomial terms exhibit a natural power structure, thereby allowing for an elegant and systematic approach to factorization.

Each of these techniques—grouping, pattern recognition, and substitution—provides robust strategies for decomposing higher-degree polynomials into products of simpler factors. Their proper application depends on the structure of the polynomial and the ability to discern underlying patterns and common factors within the expression.

Multiple Choice Questions

1. Which of the following best describes the method of factoring by grouping?

    (a) It involves finding the greatest common factor across all terms.

    (b) It partitions the polynomial into groups that share a common binomial factor.

    (c) It rewrites the polynomial as a perfect square trinomial.

    (d) It applies the quadratic formula to determine factors.

2. In the polynomial
$$x^3 + 2x^2 - x - 2,$$
    what is the most natural grouping for factoring by grouping?

    (a) $(x^3 - x) + (2x^2 - 2)$
    (b) $(x^3 + 2x^2) + (-x - 2)$
    (c) $(x^3 - 2) + (2x^2 - x)$
    (d) $(2x^2 - x) + (x^3 - 2)$

3. When factoring the expression
$$x^4 + 5x^2 + 6,$$
which substitution is most appropriate to transform it into a quadratic equation?

   (a) Let $u = x$
   (b) Let $u = x^4$
   (c) Let $u = x^2$
   (d) Let $u = 5x^2$

4. Factoring by substitution is especially useful when the exponents of the terms have a common multiple. Which substitution should you use to simplify
$$x^6 + 7x^3 + 12 \ ?$$

   (a) Let $u = x^2$
   (b) Let $u = x^3$
   (c) Let $u = x^6$
   (d) Let $u = 7x^3$

5. After applying factoring by grouping to the polynomial
$$x^3 + 2x^2 - x - 2,$$
you obtain
$$(x+2)(x^2-1).$$
What is the next step in fully factoring the expression?

   (a) Factor $x^2 - 1$ as $(x-1)(x+1)$.
   (b) Factor $x^2 - 1$ as $(x-1)^2$.
   (c) Factor $x^2 - 1$ as $(x+1)^2$.
   (d) No further factorization is possible.

6. Consider the biquadratic polynomial
$$x^4 - 5x^2 + 4.$$
Which sequence of steps correctly describes its factorization using substitution?

(a) Substitute $u = x^2$, factor as $(u-1)(u-4)$, then substitute back and factor any differences of squares.

(b) Factor directly as $(x^2 - 5)(x^2 + 4)$.

(c) Group the terms as $(x^4 - 5x^2) + 4$ and then factor.

(d) Rewrite as a perfect square and then take the square root.

7. Which of the following statements regarding pattern recognition in polynomial factoring is TRUE?

(a) Every polynomial with terms in descending order factors into linear factors.

(b) Recognizing a difference of squares within an expression can lead to further factorization.

(c) Only even-degree polynomials can be factored using substitution.

(d) Factoring by grouping applies only when a perfect square trinomial is present.

Answers:

1. **B:** Factoring by grouping involves dividing the polynomial into smaller groups in which each group has a common factor that can be factored out; this often reveals a common binomial factor between the groups.

2. **B:** Grouping the terms as $(x^3 + 2x^2) + (-x - 2)$ is natural because it allows you to factor each group separately—factoring x^2 from the first group and -1 from the second—thereby yielding the common factor $(x + 2)$.

3. **C:** Setting $u = x^2$ turns the expression $x^4 + 5x^2 + 6$ into $u^2 + 5u + 6$, a standard quadratic that can be factored easily before substituting back the original variable.

4. **B:** In the polynomial $x^6 + 7x^3 + 12$, the exponents are multiples of 3, making the substitution $u = x^3$ the most effective way to convert it into a quadratic form $u^2 + 7u + 12$.

5. **A:** The obtained factor $x^2 - 1$ is a classic difference of squares, which factors further into $(x-1)(x+1)$; this completes the full factorization of the original expression.

6. **A:** By substituting $u = x^2$, the biquadratic polynomial $x^4 - 5x^2 + 4$ becomes the quadratic $u^2 - 5u + 4$, which factors to $(u-1)(u-4)$. Re-substituting x^2 for u yields factors that can further be expressed as differences of squares.

7. **B:** Recognizing patterns such as a difference of squares is a key strategy in polynomial factoring; identifying these familiar structures simplifies the factorization process considerably.

Practice Problems

1. Factor by grouping the polynomial:
$$x^3 + 2x^2 - x - 2.$$

2. Factor the polynomial by substitution:
$$x^4 + 5x^2 + 6.$$

3. Factor the polynomial using substitution:
$$x^6 + 7x^3 + 12.$$

4. Factor completely the polynomial by first using substitution and then applying the difference of squares:
$$x^4 - 5x^2 + 4.$$

5. Factor the polynomial using the difference of squares:
$$a^4 - b^4.$$

6. Factor by grouping the polynomial:
$$2x^3 - 5x^2 - 2x + 5.$$

Answers

1. **Problem:** Factor
$$x^3 + 2x^2 - x - 2.$$
Solution: First, group the terms as follows:
$$(x^3 + 2x^2) + (-x - 2).$$
Factor out the common factors in each group:
$$x^2(x + 2) - 1(x + 2).$$
Notice that the binomial $(x + 2)$ is common, so factor it out:
$$(x + 2)(x^2 - 1).$$
Recognize that $x^2 - 1$ is a difference of squares:
$$x^2 - 1 = (x - 1)(x + 1).$$
Thus, the complete factorization is:
$$(x + 2)(x - 1)(x + 1).$$
Explanation: The polynomial was grouped into two pairs from which the common binomial $(x+2)$ was extracted. Factoring the difference of squares $x^2 - 1$ then yielded the final product of three linear factors.

2. **Problem:** Factor
$$x^4 + 5x^2 + 6.$$

   Solution: Recognize that the polynomial can be viewed as quadratic in form. Let:
$$u = x^2.$$
   Then, the expression becomes:
$$u^2 + 5u + 6.$$
   Factor the quadratic:
$$u^2 + 5u + 6 = (u + 2)(u + 3).$$
   Substitute back $u = x^2$:
$$(x^2 + 2)(x^2 + 3).$$

   Explanation: By substituting $u = x^2$, the fourth-degree polynomial is reduced to a quadratic equation, which factors into $(u + 2)(u + 3)$. Reintroducing x^2 gives the final factorization.

3. **Problem:** Factor
$$x^6 + 7x^3 + 12.$$

   Solution: Notice that the exponents are multiples of three. Let:
$$u = x^3.$$
   The polynomial becomes:
$$u^2 + 7u + 12.$$
   Factor the quadratic:
$$u^2 + 7u + 12 = (u + 3)(u + 4).$$
   After substituting back $u = x^3$, the factorization is:
$$(x^3 + 3)(x^3 + 4).$$

   Explanation: The substitution reduced the sixth-degree polynomial to a quadratic in u. Once factored, setting u back to x^3 yields the final factorization.

4. **Problem:** Factor
$$x^4 - 5x^2 + 4.$$

   Solution: Begin by letting:
$$u = x^2.$$

   The polynomial becomes:
$$u^2 - 5u + 4.$$

   Factor the quadratic:
$$u^2 - 5u + 4 = (u-1)(u-4).$$

   Substitute back $u = x^2$ to obtain:
$$(x^2 - 1)(x^2 - 4).$$

   Both $x^2 - 1$ and $x^2 - 4$ are differences of squares and factor further:
$$x^2 - 1 = (x-1)(x+1) \quad \text{and} \quad x^2 - 4 = (x-2)(x+2).$$

   Thus, the complete factorization is:
$$(x-1)(x+1)(x-2)(x+2).$$

   Explanation: The substitution $u = x^2$ transforms the quartic polynomial into a quadratic, which factors easily. Re-substitution yields two quadratic factors that are differences of squares, leading to complete factorization into linear factors.

5. **Problem:** Factor
$$a^4 - b^4.$$

   Solution: Recognize the expression as a difference of squares:
$$a^4 - b^4 = (a^2)^2 - (b^2)^2.$$

   Apply the difference of squares formula:
$$a^4 - b^4 = (a^2 - b^2)(a^2 + b^2).$$

   Notice that $a^2 - b^2$ is also a difference of squares:
$$a^2 - b^2 = (a-b)(a+b).$$

Thus, the complete factorization is:
$$(a - b)(a + b)(a^2 + b^2).$$

Explanation: The formula for the difference of squares was applied twice—first to express $a^4 - b^4$ as a product of two quadratics, and then to factor $a^2 - b^2$ further into two linear factors.

6. **Problem:** Factor
$$2x^3 - 5x^2 - 2x + 5.$$

Solution: Group the terms:
$$(2x^3 - 5x^2) + (-2x + 5).$$

Factor out the common factors from each group:
$$x^2(2x - 5) - 1(2x - 5).$$

The common binomial factor $(2x - 5)$ can be factored out:
$$(2x - 5)(x^2 - 1).$$

Recognize that $x^2 - 1$ is a difference of squares:
$$x^2 - 1 = (x - 1)(x + 1).$$

Therefore, the complete factorization is:
$$(2x - 5)(x - 1)(x + 1).$$

Explanation: Grouping the terms revealed a common binomial, which, when factored out, left a difference of squares. Factoring the difference of squares provided the final expression as a product of three linear factors.

Chapter 25

Division of Polynomials

Polynomial Long Division

Polynomial long division is a systematic procedure for dividing one polynomial by another, analogous to the long division process used with numbers. In this method, the dividend and divisor are written in descending order of degree. The process produces a quotient polynomial and a remainder, where the division algorithm guarantees that

$$\text{Dividend} = (\text{Divisor}) \cdot (\text{Quotient}) + \text{Remainder},$$

with the degree of the remainder either zero or lower than the degree of the divisor.

In the long division process, the leading term of the dividend is divided by the leading term of the divisor to generate the first term of the quotient. Multiplying the entire divisor by this term and subtracting the result from the dividend produces a new polynomial. This new polynomial serves as the updated dividend for the next step, and the process is repeated until the remaining polynomial has a degree less than that of the divisor.

1 Detailed Process and Example

Consider the division of

$$2x^3 + 3x^2 - x - 2$$

by

$$x - 1.$$

The procedure is as follows:

1. Divide the first term of the dividend, $2x^3$, by the leading term of the divisor, x, which yields $2x^2$. This becomes the first term of the quotient.

2. Multiply the entire divisor by $2x^2$:
$$2x^2 \cdot (x - 1) = 2x^3 - 2x^2.$$

3. Subtract the product from the first two terms of the dividend:
$$(2x^3 + 3x^2) - (2x^3 - 2x^2) = 5x^2.$$

4. Bring down the next term, $-x$, resulting in the new expression:
$$5x^2 - x.$$

5. Divide $5x^2$ by x to obtain $5x$, the next term in the quotient.

6. Multiply the divisor by $5x$:
$$5x \cdot (x - 1) = 5x^2 - 5x.$$

7. Subtract this product from the current dividend:
$$(5x^2 - x) - (5x^2 - 5x) = 4x.$$

8. Bring down the final term, -2, forming:
$$4x - 2.$$

9. Divide $4x$ by x to obtain 4, and multiply the divisor by 4:
$$4 \cdot (x - 1) = 4x - 4.$$

10. Subtract the product:
$$(4x - 2) - (4x - 4) = 2.$$

The quotient obtained from this process is
$$2x^2 + 5x + 4,$$
and the remainder is
$$2.$$
This result may be expressed in the division algorithm form as:
$$2x^3 + 3x^2 - x - 2 = (x - 1)(2x^2 + 5x + 4) + 2.$$

Synthetic Division

Synthetic division offers an efficient alternative to polynomial long division when the divisor is a linear binomial of the form $x - c$. This method streamlines the process by concentrating solely on the coefficients of the dividend polynomial and reducing the often cumbersome algebraic manipulations.

The procedure in synthetic division begins by writing the coefficients of the dividend in descending order of degree, inserting zeros for any missing terms. The value c is determined by setting the divisor equal to zero; that is, for $x - c = 0$, the value is c. The algorithm then proceeds with the following steps:

1. Place the value c to the left of the coefficients.

2. Bring down the first coefficient unchanged as the starting value of the quotient.

3. Multiply c by the value just brought down, and write the result beneath the next coefficient.

4. Add the column, and write the sum in the bottom row.

5. Repeat the multiplication and addition for each coefficient. The final number in the bottom row represents the remainder, while the other numbers form the coefficients of the quotient polynomial.

1 Worked Example

Examine the division of

$$x^3 - 6x^2 + 11x - 6$$

by

$$x - 1.$$

The divisor $x-1$ implies that $c = 1$. The coefficients of the dividend are 1 (for x^3), -6 (for x^2), 11 (for x), and -6 (the constant term).

The synthetic division setup is as follows:

```
1 | 1   -6   11   -6
  |      1   -5    6
  |_____
    1   -5    6    0
```

The process involves these steps:

1. Bring down the first coefficient (1).

2. Multiply 1 (the value just written) by 1 (the value of c) to get 1. Write this under the second coefficient (−6).

3. Add −6 and 1 to obtain −5.

4. Multiply −5 by 1 to get −5; write this under the third coefficient (11).

5. Adding 11 and −5 yields 6.

6. Multiply 6 by 1 to obtain 6; write this under the final coefficient (−6).

7. Adding −6 and 6 results in a remainder of 0.

The bottom row (excluding the remainder) represents the coefficients of the quotient polynomial, which is

$$x^2 - 5x + 6.$$

The zero remainder confirms that $x - 1$ is a factor of $x^3 - 6x^2 + 11x - 6$.

Synthetic division, by focusing on coefficients and simple arithmetic operations, serves as a concise method for handling the division of polynomials when the divisor is a linear expression.

Multiple Choice Questions

1. Which of the following expressions correctly represents the division algorithm for polynomials?

   (a) Dividend = Divisor + Quotient + Remainder

   (b) Dividend = Divisor × Quotient + Remainder

   (c) Dividend = Divisor - Quotient + Remainder

   (d) Dividend = Divisor × Quotient - Remainder

2. In polynomial long division, what condition must the remainder satisfy?

   (a) Its degree is greater than the degree of the divisor.

   (b) Its degree is equal to the degree of the dividend.

   (c) Its degree is less than the degree of the divisor.

(d) Its degree is exactly one less than the degree of the divisor.

3. Synthetic division is most appropriately used when the divisor is:

    (a) A quadratic polynomial.
    (b) A linear binomial of the form $x - c$.
    (c) A binomial with a leading coefficient other than 1.
    (d) Any polynomial as long as it is written in descending order.

4. In the polynomial long division process, after dividing the leading term of the dividend by the leading term of the divisor, what is the next step?

    (a) Multiply the divisor by the obtained term and add the result to the dividend.
    (b) Multiply the divisor by the obtained term and subtract the result from the dividend.
    (c) Multiply the quotient term by itself.
    (d) Rearrange the remaining terms of the dividend in ascending order.

5. In synthetic division, what does the final number in the bottom row represent?

    (a) The degree of the quotient.
    (b) The leading coefficient of the divisor.
    (c) The remainder of the division.
    (d) The sum of all the coefficients.

6. What is one significant advantage of synthetic division over long division?

    (a) It can be used with divisors of any degree.
    (b) It involves fewer steps by focusing only on the coefficients.
    (c) It automatically rearranges the polynomial in descending order.
    (d) It provides the complete factorization of the dividend.

7. When dividing $2x^3 + 3x^2 - x - 2$ by $x - 1$ using long division, what is the remainder?

   (a) -2

   (b) 0

   (c) 2

   (d) 1

Answers:

1. **B: Dividend = Divisor × Quotient + Remainder**
   This expression embodies the division algorithm for polynomials, which states that any dividend can be represented as the product of the divisor and quotient, plus a remainder.

2. **C: Its degree is less than the degree of the divisor**
   The division algorithm guarantees that the remainder will have a degree smaller than that of the divisor, ensuring that the division process is complete.

3. **B: A linear binomial of the form $x - c$**
   Synthetic division is specifically designed to work with divisors that are linear and have the form $x - c$, allowing the process to work solely with the coefficients.

4. **B: Multiply the divisor by the obtained term and subtract the result from the dividend**
   After dividing the leading term of the dividend by the leading term of the divisor, the next step is to multiply the divisor by the obtained quotient term and subtract that product from the dividend to form a new polynomial.

5. **C: The remainder of the division**
   In the synthetic division process, the final number placed in the bottom row represents the remainder after all the coefficient computations have been performed.

6. **B: It involves fewer steps by focusing only on the coefficients**
   One of the main advantages of synthetic division is its efficiency; by bypassing variable notation and concentrating on the coefficients, the process is faster and less prone to algebraic errors when the divisor is linear.

7. **C: 2**

   As shown in the worked example, dividing $2x^3 + 3x^2 - x - 2$ by $x - 1$ yields a quotient of $2x^2 + 5x + 4$ and a remainder of 2.

Practice Problems

1. Perform polynomial long division on the following expression:

$$\frac{3x^3 - 5x^2 + 2x - 8}{x - 2}.$$

2. Use polynomial long division to divide:

$$\frac{x^4 + 2x^3 - 3x^2 + x - 5}{x + 1}.$$

3. Apply synthetic division to divide:

$$\frac{2x^3 + x^2 - 4x + 3}{x + 3}.$$

4. Use synthetic division to divide:

$$\frac{x^4 - 2x^3 + 3x^2 - 4x + 5}{x - 2}.$$

5. Explain why synthetic division is applicable only when dividing by a linear polynomial of the form

$$x - c.$$

6. The Remainder Theorem states that when a polynomial
$$P(x)$$
is divided by
$$x - c,$$
the remainder is
$$P(c).$$
Given that dividing a polynomial
$$P(x)$$
by
$$x - 3$$
results in a remainder of 7, determine
$$P(3)$$
and explain your reasoning.

Answers

1. **Problem:** Divide
$$3x^3 - 5x^2 + 2x - 8$$
by
$$x - 2.$$
Solution:
Write the problem as a division:

Dividend: $3x^3 - 5x^2 + 2x - 8$,   Divisor: $x - 2$.

Step 1: Divide the leading term of the dividend by the leading term of the divisor:
$$\frac{3x^3}{x} = 3x^2.$$
This becomes the first term of the quotient.

Step 2: Multiply the divisor by $3x^2$:
$$3x^2(x - 2) = 3x^3 - 6x^2.$$

Step 3: Subtract the result from the first two terms of the dividend:
$$(3x^3 - 5x^2) - (3x^3 - 6x^2) = x^2.$$
Bring down the next term $+2x$ to get:
$$x^2 + 2x.$$

Step 4: Divide the leading term x^2 by x:
$$\frac{x^2}{x} = x.$$

Step 5: Multiply the divisor by x:
$$x(x - 2) = x^2 - 2x.$$

Step 6: Subtract:
$$(x^2 + 2x) - (x^2 - 2x) = 4x.$$
Bring down the next term -8 to form:
$$4x - 8.$$

Step 7: Divide $4x$ by x:
$$\frac{4x}{x} = 4.$$

Step 8: Multiply the divisor by 4:
$$4(x - 2) = 4x - 8.$$

Step 9: Subtract:
$$(4x - 8) - (4x - 8) = 0.$$

The quotient of the division is:
$$3x^2 + x + 4,$$

and the remainder is 0. Thus,
$$3x^3 - 5x^2 + 2x - 8 = (x - 2)(3x^2 + x + 4).$$

2. **Problem:** Divide
$$x^4 + 2x^3 - 3x^2 + x - 5$$
by
$$x + 1$$
using polynomial long division.

Solution:
Write the dividend and divisor:
$$\text{Dividend: } x^4 + 2x^3 - 3x^2 + x - 5, \quad \text{Divisor: } x + 1.$$

Step 1: Divide the leading term x^4 by x:
$$\frac{x^4}{x} = x^3.$$

Step 2: Multiply the divisor by x^3:
$$x^3(x + 1) = x^4 + x^3.$$

Step 3: Subtract:
$$(x^4 + 2x^3) - (x^4 + x^3) = x^3.$$

Bring down the next term to obtain:
$$x^3 - 3x^2.$$

Step 4: Divide x^3 by x:
$$\frac{x^3}{x} = x^2.$$

Step 5: Multiply the divisor by x^2:
$$x^2(x+1) = x^3 + x^2.$$

Step 6: Subtract:
$$(x^3 - 3x^2) - (x^3 + x^2) = -4x^2.$$

Bring down the term $+x$ to form:
$$-4x^2 + x.$$

Step 7: Divide $-4x^2$ by x:
$$\frac{-4x^2}{x} = -4x.$$

Step 8: Multiply the divisor by $-4x$:
$$-4x(x+1) = -4x^2 - 4x.$$

Step 9: Subtract:
$$(-4x^2 + x) - (-4x^2 - 4x) = 5x.$$

Bring down the last term -5 to get:
$$5x - 5.$$

Step 10: Divide $5x$ by x:
$$\frac{5x}{x} = 5.$$

Step 11: Multiply:
$$5(x+1) = 5x + 5.$$

Step 12: Subtract:
$$(5x - 5) - (5x + 5) = -10.$$

The quotient is:
$$x^3 + x^2 - 4x + 5,$$
with a remainder of -10. Therefore,
$$x^4 + 2x^3 - 3x^2 + x - 5 = (x+1)(x^3 + x^2 - 4x + 5) - 10.$$

3. **Problem:** Use synthetic division to divide
$$2x^3 + x^2 - 4x + 3$$
by
$$x + 3.$$

Solution:

First, rewrite the divisor in the form
$$x - c.$$
Since
$$x + 3 = x - (-3),$$
we have $c = -3$.

List the coefficients of the dividend:
$$2 \quad 1 \quad -4 \quad 3.$$

Set up synthetic division:

$$\begin{array}{r|rrrr} -3 & 2 & 1 & -4 & 3 \\ & & -6 & 15 & -33 \\ \hline & 2 & -5 & 11 & -30 \end{array}$$

Explanation:

(a) Bring down the first coefficient: 2.
(b) Multiply 2 by -3 to get -6; add to 1 to obtain -5.
(c) Multiply -5 by -3 to get 15; add to -4 to obtain 11.
(d) Multiply 11 by -3 to get -33; add to 3 to yield a remainder of -30.

The quotient polynomial has coefficients corresponding to:
$$2x^2 - 5x + 11,$$
and the remainder is -30. Thus,
$$2x^3 + x^2 - 4x + 3 = (x+3)(2x^2 - 5x + 11) - 30.$$

4. **Problem:** Use synthetic division to divide
$$x^4 - 2x^3 + 3x^2 - 4x + 5$$
by
$$x - 2.$$

Solution:

Since the divisor is
$$x - 2,$$
we have $c = 2$.

Write the coefficients of the dividend:
$$1 \quad -2 \quad 3 \quad -4 \quad 5.$$

Set up synthetic division:

$$\begin{array}{r|rrrrr} 2 & 1 & -2 & 3 & -4 & 5 \\ & & 2 & 0 & 6 & 4 \\ \hline & 1 & 0 & 3 & 2 & 9 \end{array}$$

Explanation:

(a) Bring down the first coefficient: 1.
(b) Multiply 1 by 2 to obtain 2; add to -2 to get 0.
(c) Multiply 0 by 2 to get 0; add to 3 to get 3.
(d) Multiply 3 by 2 to obtain 6; add to -4 to get 2.
(e) Multiply 2 by 2 to obtain 4; add to 5 to get a remainder of 9.

The quotient polynomial is:
$$x^3 + 0x^2 + 3x + 2,$$
which simplifies to:
$$x^3 + 3x + 2,$$
and the remainder is 9. Therefore,
$$x^4 - 2x^3 + 3x^2 - 4x + 5 = (x - 2)(x^3 + 3x + 2) + 9.$$

5. **Problem:** Explain why synthetic division is applicable only when dividing by a linear polynomial of the form
$$x - c.$$

   Solution:

   Synthetic division is a simplified method that focuses solely on the coefficients of a polynomial. It relies on representing the divisor in the form
   $$x - c,$$
   where the leading coefficient of x is 1. This setup allows the process to use only arithmetic operations (multiplication and addition) and omits variable manipulation. The method mirrors the Remainder Theorem, which connects the remainder of the division to the evaluation of the polynomial at c (i.e., $P(c)$). If the divisor were of higher degree or had a leading coefficient other than 1, the procedure would require additional adjustments and could no longer be performed by this streamlined process. Hence, synthetic division is directly applicable only when the divisor is a linear binomial in the form
   $$x - c.$$

6. **Problem:** The Remainder Theorem states that when a polynomial
   $$P(x)$$
   is divided by
   $$x - c,$$
   the remainder is
   $$P(c).$$
   Given that dividing a polynomial
   $$P(x)$$
   by
   $$x - 3$$
   results in a remainder of 7, determine
   $$P(3)$$
   and explain your reasoning.

282

Solution:

According to the Remainder Theorem, the remainder when a polynomial $P(x)$ is divided by $x - c$ equals $P(c)$. In this problem, the divisor is $x - 3$, which indicates $c = 3$. Therefore, the remainder of the division is exactly the value of the polynomial at $x = 3$. Given that the remainder is 7, we conclude:
$$P(3) = 7.$$

This relationship arises because the division representation of $P(x)$ is:
$$P(x) = (x - 3)Q(x) + P(3),$$
and substituting $x = 3$ causes the term $(x-3)Q(x)$ to vanish, leaving only the remainder $P(3)$.

Chapter 26

Rational Expressions: Concepts and Simplification

Definition and Fundamental Properties

A rational expression is an algebraic fraction in which both the numerator and the denominator are polynomials. In its general form, a rational expression is written as

$$\frac{P(x)}{Q(x)},$$

where $P(x)$ and $Q(x)$ are polynomials and the condition $Q(x) \neq 0$ is imposed to ensure that the expression is defined. This form embodies the concept of division in algebra and provides a framework for understanding many of the key properties within the subject. The polynomial in the numerator, $P(x)$, may be of any degree, while the denominator $Q(x)$ must be nonzero for all values of x in the domain of the expression. Such a restriction is necessary because division by zero is undefined in the real number system. The structural nature of rational expressions requires attention to both the form of the polynomials involved and the operations that can be applied to them.

Domain Restrictions and Excluded Values

In any rational expression, the values of the variable that make the denominator equal to zero must be explicitly excluded from the domain. For a rational expression given by

$$\frac{P(x)}{Q(x)},$$

the condition $Q(x) \neq 0$ is fundamental. Determining the values for which the denominator vanishes involves solving the equation

$$Q(x) = 0.$$

For example, if the denominator is expressed as

$$x^2 - 4,$$

it can be factored into

$$(x-2)(x+2) = 0.$$

Solving for x gives the solutions $x = 2$ and $x = -2$. These values are then excluded from the domain of the rational expression. The identification of such restrictions is crucial, as it preserves the validity of any subsequent algebraic manipulations, particularly those involving factorization and cancellation. Even when identical factors are eliminated during simplification, the variable restrictions originally imposed by the denominator must remain in force.

Simplification Techniques Through Factoring and Cancellation

The process of simplifying a rational expression often begins with factoring both the numerator and the denominator into irreducible polynomials. Once factorization is complete, any common factors that appear in both the numerator and the denominator can be cancelled, provided that the factors represent nonzero quantities. Consider the rational expression

$$\frac{x^2 - 9}{x^2 - 6x + 9}.$$

The numerator factors into
$$(x-3)(x+3),$$
and the denominator factors into
$$(x-3)(x-3).$$
After identifying the common factor $(x-3)$, cancellation yields the simplified expression
$$\frac{x+3}{x-3}.$$
It is important to note that despite the cancellation, the restriction $x \neq 3$ must still be observed as it was inherent in the original function. This preservation of the domain restrictions is essential even when the apparent complexity of the expression decreases after simplification.

Additional examples further illustrate the power of effective factorization. Take, for instance, the rational expression
$$\frac{2x^2-8}{4x^2-16}.$$
The numerator may be factored by first extracting the common factor 2, resulting in
$$2(x^2-4),$$
and the denominator by extracting 4 to obtain
$$4(x^2-4).$$
Recognizing that the quadratic $x^2 - 4$ is a difference of squares, it factors as
$$(x-2)(x+2).$$
Thus, the expression is rewritten as
$$\frac{2(x-2)(x+2)}{4(x-2)(x+2)},$$
which simplifies to
$$\frac{2}{4} = \frac{1}{2},$$
subject to the restrictions $x \neq 2$ and $x \neq -2$ that originate from the factorization of $x^2 - 4$.

The method of cancellation through factoring provides a systematic way to simplify rational expressions while maintaining a clear account of all variable restrictions. This careful approach ensures that every algebraic manipulation remains mathematically rigorous and that the final simplified expression is equivalent to the original, subject to the same domain limitations.

Multiple Choice Questions

1. Which of the following best defines a rational expression?

    (a) An expression that represents the ratio of two numbers.

    (b) An algebraic fraction where both the numerator and denominator are polynomials and the denominator is not zero.

    (c) A fraction with only constant values in the numerator and denominator.

    (d) An expression that only involves division of monomials.

2. In the general form of a rational expression
$$\frac{P(x)}{Q(x)},$$
which condition is mandatory to ensure the expression is defined?

    (a) $P(x) \neq 0$

    (b) $Q(x) \neq 0$

    (c) $P(x) = Q(x)$

    (d) $P(x)$ and $Q(x)$ must both be factorable

3. When simplifying a rational expression by factoring and cancellation, which statement is correct?

    (a) Cancelled factors no longer impose any restrictions on the domain.

    (b) Only the numerator needs to be factored for valid cancellation.

    (c) After cancellation, the original restrictions from the denominator must still be observed.

(d) It is unnecessary to check for any common factors before cancellation.

4. The quadratic expression $x^2 - 9$ is factored using which technique?

    (a) Factoring by grouping
    (b) Factoring a perfect square trinomial
    (c) Using the difference of squares method
    (d) Factoring by extracting the greatest common factor

5. Consider the rational expression
$$\frac{x^2 - 9}{x^2 - 6x + 9}.$$
Upon factoring, what is the simplified form (assuming proper domain restrictions)?

    (a) $\frac{x-3}{x+3}$
    (b) $\frac{x+3}{x-3}$
    (c) $\frac{x-3}{x-3}$
    (d) $\frac{x+3}{x+3}$

6. In the process of simplifying the expression
$$\frac{2x^2 - 8}{4x^2 - 16},$$
what is the correct initial step when factoring the numerator $2x^2 - 8$?

    (a) Factor out x to obtain $x(2x - 8)$
    (b) Factor out 2 to obtain $2(x^2 - 4)$
    (c) Factor the quadratic directly as a difference of squares
    (d) Divide the numerator and denominator by $2x$

7. Which of the following operations is NOT valid when simplifying a rational expression?

    (a) Multiplying the numerator and denominator by the same nonzero expression.

(b) Factoring both the numerator and denominator to identify common factors.

(c) Canceling a common factor that could be zero for some value of x.

(d) Maintaining the domain restrictions even after cancellation.

Answers:

1. **B: An algebraic fraction where both the numerator and denominator are polynomials and the denominator is not zero**
   This option correctly captures the definition of a rational expression, emphasizing that both parts must be polynomials and that the denominator must not equal zero to avoid undefined expressions.

2. **B: $Q(x) \neq 0$**
   For a rational expression $\frac{P(x)}{Q(x)}$ to be defined, the denominator $Q(x)$ must not be zero. This condition is fundamental due to the undefined nature of division by zero.

3. **C: After cancellation, the original restrictions from the denominator must still be observed.**
   Even after common factors are cancelled, the values that made these factors zero in the original expression must still be excluded from the domain to maintain mathematical rigor.

4. **C: Using the difference of squares method**
   The expression $x^2 - 9$ can be viewed as $x^2 - 3^2$, which factors neatly into $(x - 3)(x + 3)$ via the difference of squares technique.

5. **B: $\frac{x+3}{x-3}$**
   Factoring the numerator gives $(x-3)(x+3)$ and the denominator factors as $(x-3)(x-3)$. Cancelling the common factor $(x-3)$ results in $\frac{x+3}{x-3}$; however, note that $x \neq 3$ must be maintained.

6. **B: Factor out 2 to obtain $2(x^2 - 4)$**
   The correct first step is to factor out the greatest common factor, which is 2. This simplifies the expression to $2(x^2 - 4)$, where the quadratic $x^2 - 4$ can further be factored as a difference of squares.

7. **C: Canceling a common factor that could be zero for some value of x.**

   Canceling a common factor is only valid if that factor is nonzero. If the factor can be zero for some x-value, then canceling it ignores necessary domain restrictions and could lead to incorrect conclusions.

Practice Problems

1. Determine the domain of the rational expression:

$$r(x) = \frac{x^2 - 4}{x^2 - 5x + 6}$$

2. Simplify the following rational expression and state the domain restrictions:

$$S(x) = \frac{x^2 - 9}{x^2 - 6x + 9}$$

3. Simplify the rational expression below by factoring and cancellation, and specify the excluded values:

$$T(x) = \frac{2x^2 - 8}{4x^2 - 16}$$

4. Factor and simplify the expression, then list the domain restrictions:

$$U(x) = \frac{x^2 - 4}{x^2 - x - 6}$$

5. Simplify the following rational expression by canceling common factors, and determine the restrictions on the variable:

$$V(x) = \frac{3x^2 - 12x}{x^2 - 4x}$$

6. Explain why it is necessary to state the domain restrictions of a rational expression even after cancellation of common factors.

Answers

1. **Solution:** The domain of a rational expression consists of all real numbers except those that make the denominator zero. First, factor the denominator:
$$x^2 - 5x + 6 = (x-2)(x-3).$$
Setting each factor equal to zero gives $x - 2 = 0$ or $x - 3 = 0$, which yields $x = 2$ and $x = 3$. Therefore, the domain of $r(x)$ is all real numbers except $x = 2$ and $x = 3$.

2. **Solution:** Begin by factoring both the numerator and the denominator.
$$x^2 - 9 = (x-3)(x+3),$$
$$x^2 - 6x + 9 = (x-3)^2.$$
Cancel the common factor $(x-3)$ noting that $x \neq 3$ (since this would make the original denominator zero). The simplified expression becomes:
$$S(x) = \frac{x+3}{x-3}.$$
The domain restriction is that x must not equal 3, so the expression is defined for all $x \neq 3$.

3. **Solution:** Factor both the numerator and the denominator:
$$2x^2 - 8 = 2(x^2 - 4) = 2(x-2)(x+2),$$

292

$$4x^2 - 16 = 4(x^2 - 4) = 4(x-2)(x+2).$$

Cancel the common factors $(x-2)(x+2)$ (with the understanding that $x \neq 2$ and $x \neq -2$ because these values make the factors zero). The expression reduces to:

$$T(x) = \frac{2}{4} = \frac{1}{2}.$$

So, the simplified expression is $\frac{1}{2}$ with the restrictions $x \neq 2$ and $x \neq -2$.

4. **Solution:** Factor the numerator and the denominator:

$$x^2 - 4 = (x-2)(x+2),$$

$$x^2 - x - 6 = (x-3)(x+2).$$

Cancel the common factor $(x+2)$, with the condition that $x \neq -2$ (to avoid division by zero). The simplified expression is:

$$U(x) = \frac{x-2}{x-3}.$$

From the original denominator, we also require that $x \neq 3$. Therefore, the simplified form is valid provided $x \neq -2$ and $x \neq 3$.

5. **Solution:** Factor the numerator and the denominator:

$$3x^2 - 12x = 3x(x-4),$$

$$x^2 - 4x = x(x-4).$$

Cancel the common factors x and $(x-4)$, making sure that $x \neq 0$ and $x \neq 4$ (since these values would make the factors zero). The expression simplifies to:

$$V(x) = 3.$$

Hence, the simplified expression is 3 with the restrictions $x \neq 0$ and $x \neq 4$.

6. **Explanation:** When simplifying rational expressions, cancellation of common factors is only valid if the factors being cancelled are nonzero. The original expression may have restrictions (values that make the denominator zero) that persist even after cancellation. If we cancel a factor without

noting these restrictions, we might incorrectly assume that the simplified expression is defined for all values of x. However, the cancelled factor represents a condition that must be met to avoid division by zero. Therefore, it is essential to state these domain restrictions to ensure that the simplified expression is equivalent to the original expression over the correct domain.

Chapter 27

Operations with Rational Expressions

Addition and Subtraction of Rational Expressions

The addition and subtraction of rational expressions require the consolidation of different denominators into a single, common denominator. This process begins with factoring all denominators into their irreducible polynomial components. The least common denominator (LCD) is then determined by taking each distinct factor to the highest power with which it appears in any of the denominators. Once the LCD is established, each rational expression is rewritten with this common denominator. The numerators are subsequently combined through either addition or subtraction, and the resulting expression is simplified by distributing terms and combining like terms.

For example, given two rational expressions

$$\frac{P(x)}{(x-a)(x-b)} \quad \text{and} \quad \frac{Q(x)}{(x-a)(x-c)},$$

their LCD is

$$(x-a)(x-b)(x-c).$$

Each expression is rewritten by multiplying the numerator and denominator by the appropriate factor, so that both fractions share the LCD. The combined expression can then be written as

$$\frac{P(x)(x-c) \pm Q(x)(x-b)}{(x-a)(x-b)(x-c)},$$

with the algebraic operations performed on the numerators followed by any applicable factorization and cancellation of common factors. It is essential to retain any domain restrictions that arise from the factors in the denominators.

Multiplication of Rational Expressions

Multiplication of rational expressions is performed by multiplying the numerators together to form a new numerator, and multiplying the denominators together to form a new denominator. Prior to executing the multiplication, it is advantageous to factor each numerator and denominator completely. Such factorization facilitates the cancellation of any common factors between the numerators and denominators, thereby reducing the expression to a simpler form.

Consider two rational expressions in factored form:

$$\frac{A(x)}{B(x)} \quad \text{and} \quad \frac{C(x)}{D(x)}.$$

Their product is given by

$$\frac{A(x) \cdot C(x)}{B(x) \cdot D(x)}.$$

After multiplication, any common factors present in both $A(x) \cdot C(x)$ and $B(x) \cdot D(x)$ should be cancelled, provided that these factors do not equal zero. This cancellation must be performed with care to ensure that all domain restrictions stipulated by the original denominators remain in force. The systematic approach of factoring, multiplying, and then cancelling common factors produces a simplified product that retains the mathematical equivalence to the original expression.

Division of Rational Expressions

Division of rational expressions is handled by transforming the division operation into multiplication via the reciprocal of the divisor.

Given a dividend in the form
$$\frac{P(x)}{Q(x)}$$
and a divisor in the form
$$\frac{R(x)}{S(x)},$$
the division is rewritten as the multiplication
$$\frac{P(x)}{Q(x)} \cdot \frac{S(x)}{R(x)}.$$

Prior to multiplication, both the dividend and the reciprocal of the divisor should be factored thoroughly. Factoring ensures the identification and cancellation of any common factors between the numerators and denominators. For instance, after expressing all polynomials in their factored forms, any factor common to both the numerator and the denominator is cancelled, under the condition that the factor is nonzero.

The procedure relies on the prerequisite that the divisor $\frac{R(x)}{S(x)}$ is not equal to zero, as the operation of taking its reciprocal is only valid under this condition. Domain restrictions imposed by both the original denominator $Q(x)$ and the factor $R(x)$ must be carefully observed through the division process. The final expression, after all cancellations and simplifications have been performed, remains equivalent to the original division expression, subject to the same restrictions that guarantee the expression is defined.

Multiple Choice Questions

1. When adding or subtracting rational expressions, what is the first step you should take?

    (a) Multiply the numerators.
    (b) Factor all denominators into irreducible components.
    (c) Cancel any common factors in the numerators.
    (d) Multiply both the numerator and the denominator by the opposite expression.

2. In order to combine rational expressions with different denominators, the least common denominator (LCD) is found by:

(a) Multiplying the denominators together without modification.

(b) Adding the denominators.

(c) Taking each distinct factor to the highest power with which it appears in any denominator.

(d) Using only the common factors present in every denominator.

3. When multiplying two rational expressions, which of the following is the best practice before performing the multiplication?

(a) Convert the division to multiplication.

(b) Factor all numerators and denominators completely.

(c) Rewrite each expression with a common denominator.

(d) Cross-multiply the expressions.

4. What is the main purpose of canceling common factors when simplifying the product of two rational expressions?

(a) To change the original expression into a completely different one.

(b) To reduce the expression to its simplest form while preserving domain restrictions.

(c) To eliminate the need for a least common denominator.

(d) To ensure that all factors in the numerator are removed.

5. In dividing rational expressions, why do we replace division with multiplication by the reciprocal of the divisor?

(a) Because it simplifies the process by converting the division into multiplication, allowing for factoring and cancellation.

(b) Because the reciprocal always cancels the numerator.

(c) Because it eliminates any need for factoring.

(d) Because it combines the numerators of both expressions.

6. After rewriting a division of rational expressions as multiplication by the reciprocal, what step should you perform next?

(a) Multiply the numerators and denominators without any further modifications.

(b) Factor all expressions completely and cancel any common factors.

(c) Add the numerators together.

(d) Expand all products fully before simplifying.

7. Throughout operations on rational expressions, why is it essential to consider domain restrictions?

(a) To confirm that the expressions are defined by excluding any values that would make a denominator zero.

(b) To determine the degrees of the polynomials involved.

(c) To allow cancellation of factors regardless of extraneous solutions.

(d) To facilitate finding a common denominator.

Answers:

1. **B: Factor all denominators into irreducible components**
Explanation: Factoring the denominators first helps in identifying all the factors, which is essential for determining the LCD needed to combine the expressions correctly.

2. **C: Taking each distinct factor to the highest power with which it appears in any denominator**
Explanation: The LCD must include every unique factor from all denominators at their highest power to ensure that each original denominator divides the LCD evenly.

3. **B: Factor all numerators and denominators completely**
Explanation: Complete factorization before multiplying allows you to cancel any common factors, simplifying the final expression and reducing computational complexity.

4. **B: To reduce the expression to its simplest form while preserving domain restrictions**
Explanation: Canceling common factors simplifies the expression without altering its value, but one must always keep the original restrictions (values that make any denominator zero) in mind.

5. **A: Because it simplifies the process by converting the division into multiplication, allowing for factoring and cancellation**

Explanation: Rewriting division as multiplication by the reciprocal standardizes the process, enabling the use of familiar techniques like factoring and canceling common factors.

6. **B: Factor all expressions completely and cancel any common factors**
   Explanation: After rewriting the division as multiplication, factoring is crucial to simplify the expression by eliminating common factors, which streamlines the result without violating any domain restrictions.

7. **A: To confirm that the expressions are defined by excluding any values that would make a denominator zero**
   Explanation: Domain restrictions are critical because they ensure that the expression remains valid; any value that makes a denominator equal to zero must be excluded, even if that factor cancels during simplification.

Practice Problems

1. Combine the rational expressions by finding a common denominator and simplifying:

$$\frac{2}{x-1} + \frac{5}{x+3}$$

2. Multiply the following rational expressions and simplify the result:

$$\frac{x^2-4}{x+2} \cdot \frac{x+2}{x-2}$$

3. Divide the following rational expressions and simplify, ensuring to note any restrictions:

$$\frac{x^2-9}{x^2-4x+3} \div \frac{x-3}{x-1}$$

4. Subtract and simplify the rational expressions below:

$$\frac{3}{x^2-4} - \frac{2}{x-2}$$

5. Simplify the expression by combining terms over a common denominator:

$$\frac{x}{x^2-1} + \frac{2}{x+1}$$

6. Simplify the following complex rational expression completely, and state the domain restrictions:
$$\frac{2(x^2-4)}{x^2-3x-4} \div \frac{x^2-9}{x^2-4}$$

Answers

1. **Solution:**
   The denominators are $x-1$ and $x+3$, so the least common denominator (LCD) is
   $$(x-1)(x+3).$$
   Rewrite each fraction with the LCD:
   $$\frac{2}{x-1} = \frac{2(x+3)}{(x-1)(x+3)} \quad \text{and} \quad \frac{5}{x+3} = \frac{5(x-1)}{(x-1)(x+3)}.$$
   Now combine the fractions:
   $$\frac{2(x+3)+5(x-1)}{(x-1)(x+3)} = \frac{2x+6+5x-5}{(x-1)(x+3)} = \frac{7x+1}{(x-1)(x+3)}.$$
   Answer:
   $$\frac{7x+1}{(x-1)(x+3)} \quad \text{with } x \neq 1,\ x \neq -3.$$

2. **Solution:**
   First, factor the numerator of the first fraction:
   $$x^2 - 4 = (x-2)(x+2).$$

Substitute this into the expression:
$$\frac{(x-2)(x+2)}{x+2} \cdot \frac{x+2}{x-2}.$$

Cancel the common factor $(x+2)$ in the first fraction (provided $x \neq -2$):
$$= \frac{x-2}{1} \cdot \frac{x+2}{x-2}.$$

Then cancel the common factor $(x-2)$ (provided $x \neq 2$):
$$= \frac{x+2}{1} = x+2.$$

Answer:
$$x+2 \quad \text{with } x \neq -2,\ x \neq 2.$$

3. **Solution:**
Factor the quadratics:
$$x^2 - 9 = (x-3)(x+3) \quad \text{and} \quad x^2 - 4x + 3 = (x-1)(x-3).$$

Rewrite the division problem:
$$\frac{(x-3)(x+3)}{(x-1)(x-3)} \div \frac{x-3}{x-1}.$$

Cancel the common factor $(x-3)$ in the first fraction (provided $x \neq 3$):
$$= \frac{x+3}{x-1} \div \frac{x-3}{x-1}.$$

Rewrite the division as multiplication by the reciprocal:
$$= \frac{x+3}{x-1} \cdot \frac{x-1}{x-3}.$$

Cancel the factor $(x-1)$ (provided $x \neq 1$):
$$= \frac{x+3}{x-3}.$$

Answer:
$$\frac{x+3}{x-3} \quad \text{with } x \neq 1,\ x \neq 3.$$

4. **Solution:**
   Recognize that:
   $$x^2 - 4 = (x-2)(x+2).$$
   Thus, the first term is:
   $$\frac{3}{(x-2)(x+2)}.$$
   The second term is:
   $$\frac{2}{x-2}.$$
   The LCD for these fractions is $(x-2)(x+2)$. Rewrite the second fraction with the LCD by multiplying numerator and denominator by $(x+2)$:
   $$\frac{2(x+2)}{(x-2)(x+2)}.$$
   Now subtract:
   $$\frac{3 - 2(x+2)}{(x-2)(x+2)} = \frac{3 - 2x - 4}{(x-2)(x+2)} = \frac{-2x - 1}{(x-2)(x+2)}.$$
   Answer:
   $$\frac{-2x - 1}{(x-2)(x+2)} \text{ with } x \neq 2,\ x \neq -2.$$

5. **Solution:**
   Note that:
   $$x^2 - 1 = (x-1)(x+1).$$
   Thus, the first fraction becomes:
   $$\frac{x}{(x-1)(x+1)}.$$
   The second fraction is:
   $$\frac{2}{x+1}.$$
   The common denominator is $(x-1)(x+1)$. Rewrite the second fraction by multiplying numerator and denominator by $(x-1)$:
   $$\frac{2(x-1)}{(x-1)(x+1)}.$$

Now, add the fractions:
$$\frac{x+2(x-1)}{(x-1)(x+1)} = \frac{x+2x-2}{(x-1)(x+1)} = \frac{3x-2}{(x-1)(x+1)}.$$

Answer:
$$\frac{3x-2}{(x-1)(x+1)} \quad \text{with } x \neq 1,\ x \neq -1.$$

6. **Solution:**
Start by factoring each quadratic:
$$x^2 - 4 = (x-2)(x+2),$$
$$x^2 - 3x - 4 = (x-4)(x+1),$$
$$x^2 - 9 = (x-3)(x+3).$$

Rewrite the original expression:
$$\frac{2(x^2-4)}{x^2-3x-4} \div \frac{x^2-9}{x^2-4} = \frac{2(x-2)(x+2)}{(x-4)(x+1)} \div \frac{(x-3)(x+3)}{(x-2)(x+2)}.$$

Convert the division to multiplication by the reciprocal:
$$= \frac{2(x-2)(x+2)}{(x-4)(x+1)} \cdot \frac{(x-2)(x+2)}{(x-3)(x+3)}.$$

Multiply the numerators and denominators:
$$= \frac{2(x-2)^2(x+2)^2}{(x-4)(x+1)(x-3)(x+3)}.$$

Answer:
$$\frac{2(x-2)^2(x+2)^2}{(x-4)(x+1)(x-3)(x+3)}$$

with the domain restrictions:
$$x \neq 2,\ x \neq -2,\ x \neq 4,\ x \neq -1,\ x \neq 3,\ x \neq -3.$$

Chapter 28

Solving Rational Equations

Identifying Domain Restrictions

A rational equation is an equation that contains one or more rational expressions, each of which is a fraction whose numerator and denominator are polynomials. It is imperative to begin the solution process by identifying all values for the variable that result in a zero denominator. For each rational expression present in the equation, the denominators must be factored completely. Values of the variable that make any of these factors equal to zero are excluded from the solution set. For instance, if an expression has a denominator of the form

$$(x-a)(x-b),$$

then the values $x = a$ and $x = b$ are not permitted. These restrictions ensure that all subsequent manipulations maintain mathematical validity.

Clearing the Denominators Using the Least Common Denominator

After determining the domain restrictions, the next step is to eliminate the fractions by multiplying both sides of the equation by the least common denominator (LCD). The LCD is obtained by factoring each of the denominators into its irreducible polynomial factors and then forming the product of all distinct factors, each raised to the highest power with which it appears in any denominator. If the rational equation contains denominators such as

$$\frac{P(x)}{(x-a)(x-b)} \quad \text{and} \quad \frac{Q(x)}{(x-a)(x-c)},$$

their LCD is given by

$$(x-a)(x-b)(x-c).$$

Multiplying every term in the equation by the LCD eliminates the fractions and yields an equation whose terms are polynomials. Care must be taken during this process to retain the original domain restrictions, as the multiplication may introduce candidate solutions that are not valid in the initial equation.

Solving the Resultant Equation

Once the fractions have been cleared, the equation transforms into a polynomial equation. The resulting equation is typically obtained by expanding products, combining like terms, and simplifying algebraic expressions. Standard techniques for solving polynomial equations are then applied. For example, if the cleared equation takes the form

$$A(x) = B(x),$$

algebraic manipulation may lead to an equation such as

$$C(x) = 0,$$

which can be solved by methods including factoring, applying the quadratic formula, or using other appropriate algebraic approaches. Each candidate solution of the simplified equation arises as a result of the algebraic manipulations performed after eliminating the denominators. It is important to execute each step with precision to ensure that all terms are correctly combined and that the integrity of the original equation is preserved.

Verification of Solutions and Identification of Extraneous Roots

The process of clearing denominators by multiplication is a powerful tool that simplifies the equation but may inadvertently introduce extraneous solutions. Such extraneous solutions are values that satisfy the polynomial equation obtained after clearing denominators but do not satisfy the original rational equation because they lead to a zero in one of the original denominators. Therefore, it is essential to substitute each candidate solution back into the original rational equation to verify that none of the denominators become zero. Only those solutions that satisfy the original restrictions are accepted. This verification step safeguards against including invalid solutions and ensures the accuracy of the final answer.

Multiple Choice Questions

1. Which of the following is the FIRST step in solving a rational equation?

    (a) Multiply every term by the least common denominator.
    (b) Identify the domain restrictions by setting the denominator(s) equal to zero.
    (c) Factor the numerators.
    (d) Simplify the entire equation by combining like terms.

2. When identifying domain restrictions in a rational equation, why is it necessary to factor the denominators completely?

    (a) To simplify the numerator.
    (b) To find all values that make any denominator zero.
    (c) To cancel common factors between the numerator and denominator.
    (d) To determine the degree of the polynomial.

3. What is the primary purpose of multiplying both sides of a rational equation by the least common denominator (LCD)?

    (a) To eliminate the fractions and obtain a polynomial equation.

(b) To identify possible extraneous solutions.

   (c) To factor the resulting expression further.

   (d) To reduce the degree of the polynomial.

4. Consider the rational expressions with denominators $(x - a)(x - b)$ and $(x - a)(x - c)$. What is the LCD of these expressions?

   (a) $(x - a)(x - b)$
   (b) $(x - a)(x - c)$
   (c) $(x - a)(x - b)(x - c)$
   (d) $(x - b)(x - c)$

5. Upon clearing the denominators, the equation transforms into a polynomial equation. Which method might you use to solve this new equation?

   (a) By applying methods such as factoring, the quadratic formula, or other algebraic techniques.

   (b) By recombining the original rational expressions.

   (c) By finding a new LCD for the polynomial.

   (d) By graphing only the denominators.

6. Why is it important to substitute candidate solutions back into the original rational equation after solving it?

   (a) To verify that the computational process was done correctly.

   (b) To ensure that none of the solutions cause any denominator to become zero.

   (c) To simplify the polynomial equation further.

   (d) To check if the numerators become zero.

7. Which of the following best describes an extraneous solution in the context of solving rational equations?

   (a) A solution that satisfies the polynomial equation obtained after clearing the denominators but does not satisfy the original rational equation.

   (b) A solution that simplifies both the numerator and the denominator.

(c) A solution that makes all terms in the equation positive.

(d) A solution that reduces the degree of the polynomial.

Answers:

1. **B: Identify the domain restrictions by setting the denominator(s) equal to zero.**
   Explanation: Before any manipulations, it is critical to determine the values for which the rational expressions are undefined—that is, where any denominator equals zero. This prevents you from accepting values that would make the original equation undefined.

2. **B: To find all values that make any denominator zero.**
   Explanation: Fully factoring the denominators reveals every factor that can potentially become zero. This ensures that all values that must be excluded from the solution set due to making a denominator zero are identified.

3. **A: To eliminate the fractions and obtain a polynomial equation.**
   Explanation: Multiplying both sides of the equation by the LCD clears the fractions, converting the rational equation into a polynomial equation that is typically easier to solve using standard algebraic techniques.

4. **C:** $(x-a)(x-b)(x-c)$
   Explanation: The LCD must contain every distinct factor from each denominator. Since one denominator has factors $(x-a)$ and $(x-b)$ and the other has $(x-a)$ and $(x-c)$, the LCD is the product of $(x-a)$, $(x-b)$, and $(x-c)$.

5. **A: By applying methods such as factoring, the quadratic formula, or other algebraic techniques.**
   Explanation: Once the denominators have been cleared, the equation becomes a standard polynomial equation. Solving it can involve factoring, applying the quadratic formula, or other algebraic methods depending on its degree and complexity.

6. **B: To ensure that none of the solutions cause any denominator to become zero.**
   Explanation: Substituting candidate solutions back into the original rational equation is essential because the process

of clearing denominators can introduce extraneous solutions that, while valid for the polynomial, make the original rational expressions undefined.

7. **A: A solution that satisfies the polynomial equation obtained after clearing the denominators but does not satisfy the original rational equation.**
   Explanation: Extraneous solutions emerge during the algebraic manipulation process (such as multiplying by the LCD) and do not meet the restrictions of the original rational equation. Verifying each candidate ensures only valid solutions are accepted.

Practice Problems

1. Consider the rational equation
$$\frac{3}{x-5} - \frac{4}{2x+1} = \frac{1}{x-5}.$$
Identify all the domain restrictions for the variable x.

2. Find the least common denominator (LCD) for the rational expressions
$$\frac{1}{(x-2)(x+3)} \quad \text{and} \quad \frac{2}{(x-2)(x-4)}.$$

3. Solve the following rational equation and identify any extraneous solutions:
$$\frac{2x}{x-1} + \frac{3}{x+2} = \frac{5x+1}{(x-1)(x+2)}.$$

4. Solve the rational equation below, showing all steps. Make sure your final answers satisfy the original equation:
$$\frac{3}{x+2} + \frac{2}{x-1} = \frac{5}{(x+2)(x-1)} + 1.$$

5. Explain, in your own words, why it is necessary to verify candidate solutions by substituting them back into the original rational equation after clearing denominators.

6. Solve the following rational equation completely. Begin by stating the domain restrictions, clear the denominators, solve the resultant equation, and then verify your solution:

$$\frac{2x+3}{x^2-4} = \frac{1}{x-2} + \frac{1}{x+2}.$$

Answers

1. **Solution:** The given equation is

$$\frac{3}{x-5} - \frac{4}{2x+1} = \frac{1}{x-5}.$$

Identify the denominators:

- In $\frac{3}{x-5}$ and $\frac{1}{x-5}$: $x - 5 \neq 0$ so $x \neq 5$.
- In $\frac{4}{2x+1}$: $2x + 1 \neq 0$ so $2x \neq -1$ and thus $x \neq -\frac{1}{2}$.

Therefore, the domain restrictions are

$$x \neq 5 \quad \text{and} \quad x \neq -\frac{1}{2}.$$

2. **Solution:** The given denominators are:

$$D_1 = (x-2)(x+3) \quad \text{and} \quad D_2 = (x-2)(x-4).$$

To find the LCD, list all distinct factors:

- The common factor is $(x-2)$.
- The remaining factors are $(x+3)$ from D_1 and $(x-4)$ from D_2.

Thus, the least common denominator is
$$\text{LCD} = (x-2)(x+3)(x-4).$$

3. **Solution:** We begin with the equation
$$\frac{2x}{x-1} + \frac{3}{x+2} = \frac{5x+1}{(x-1)(x+2)}.$$

   Step 1. Identify domain restrictions:
   - $x - 1 \neq 0 \implies x \neq 1$.
   - $x + 2 \neq 0 \implies x \neq -2$.

   Step 2. Clear the denominators: The LCD is $(x-1)(x+2)$. Multiply every term by $(x-1)(x+2)$:
   $$(x-1)(x+2) \cdot \frac{2x}{x-1} + (x-1)(x+2) \cdot \frac{3}{x+2}$$
   $$= (x-1)(x+2) \cdot \frac{5x+1}{(x-1)(x+2)}.$$

   Simplify each term:
   $$2x(x+2) + 3(x-1) = 5x + 1.$$

   Step 3. Simplify and solve: Expand the left-hand side:
   $$2x^2 + 4x + 3x - 3 = 2x^2 + 7x - 3.$$

   Set the equation:
   $$2x^2 + 7x - 3 = 5x + 1.$$

   Subtract $5x + 1$ from both sides:
   $$2x^2 + 7x - 3 - 5x - 1 = 2x^2 + 2x - 4 = 0.$$

   Divide the entire equation by 2:
   $$x^2 + x - 2 = 0.$$

   Factor the quadratic:
   $$(x+2)(x-1) = 0.$$

The candidate solutions are:
$$x = -2 \quad \text{or} \quad x = 1.$$

Step 4. Check against the domain restrictions: Both $x = -2$ and $x = 1$ are not allowed because they make the original denominators zero. *Therefore, the equation has no valid solutions.*

4. **Solution:** Consider the equation
$$\frac{3}{x+2} + \frac{2}{x-1} = \frac{5}{(x+2)(x-1)} + 1.$$

Step 1. Domain Restrictions:
- $x + 2 \neq 0 \implies x \neq -2.$
- $x - 1 \neq 0 \implies x \neq 1.$

Step 2. Clear the denominators: The LCD is $(x+2)(x-1)$. Multiply every term by $(x+2)(x-1)$:

$$(x+2)(x-1)\left[\frac{3}{x+2} + \frac{2}{x-1}\right] = (x+2)(x-1)\left[\frac{5}{(x+2)(x-1)} + 1\right].$$

Simplify term-by-term:
$$3(x-1) + 2(x+2) = 5 + (x+2)(x-1).$$

Step 3. Expand and simplify: Compute the left side:
$$3(x-1) + 2(x+2) = 3x - 3 + 2x + 4 = 5x + 1.$$

On the right side, note that
$$(x+2)(x-1) = x^2 + x - 2.$$

So, the equation becomes:
$$5x + 1 = 5 + x^2 + x - 2.$$

Simplify the right-hand side:
$$5 + x^2 + x - 2 = x^2 + x + 3.$$

Now, set the equation:
$$5x + 1 = x^2 + x + 3.$$
Rearranging the terms gives:
$$0 = x^2 + x + 3 - 5x - 1 = x^2 - 4x + 2.$$

Step 4. Solve the quadratic equation: Use the quadratic formula:
$$x = \frac{4 \pm \sqrt{(-4)^2 - 4(1)(2)}}{2} = \frac{4 \pm \sqrt{16 - 8}}{2} = \frac{4 \pm \sqrt{8}}{2}.$$
Simplify $\sqrt{8} = 2\sqrt{2}$:
$$x = \frac{4 \pm 2\sqrt{2}}{2} = 2 \pm \sqrt{2}.$$

Step 5. Verify with domain restrictions: Check that neither solution equals -2 nor 1. Since $2 + \sqrt{2} \approx 3.41$ and $2 - \sqrt{2} \approx 0.59$, both are acceptable.

Final Answer:
$$x = 2 + \sqrt{2} \quad \text{or} \quad x = 2 - \sqrt{2}.$$

5. **Solution:** When solving rational equations, it is common to multiply both sides by the least common denominator (LCD) to eliminate the fractions. However, this process may introduce extraneous solutions—values that satisfy the new, cleared equation but not the original rational equation. These extraneous solutions occur because the LCD multiplication can cancel factors that impose restrictions (i.e., values that make any original denominator zero). Verifying candidate solutions by substituting them back into the original equation ensures that none of them cause division by zero and that they truly satisfy the starting equation. This step is essential to confirm the validity of the final answers.

6. **Solution:** We start with the rational equation
$$\frac{2x+3}{x^2-4} = \frac{1}{x-2} + \frac{1}{x+2}.$$

Step 1. Domain Restrictions: Notice that
$$x^2 - 4 = (x-2)(x+2).$$
Therefore, we require:

- $x - 2 \neq 0 \implies x \neq 2,$
- $x + 2 \neq 0 \implies x \neq -2.$

Step 2. Simplify the Right-Hand Side: Combine the two fractions on the right:

$$\frac{1}{x-2} + \frac{1}{x+2} = \frac{(x+2)+(x-2)}{(x-2)(x+2)} = \frac{2x}{x^2-4}.$$

Step 3. Set Up the Equation: The equation becomes

$$\frac{2x+3}{x^2-4} = \frac{2x}{x^2-4}.$$

Since $x^2 - 4 \neq 0$ for allowable x, multiply both sides by $x^2 - 4$:

$$2x + 3 = 2x.$$

Step 4. Solve: Subtract $2x$ from both sides:

$$3 = 0.$$

This is a contradiction, which means that no value of x can satisfy the equation.

Final Answer: The equation has no valid solutions.

Chapter 29

Radical Expressions and Their Properties

Definition and Notation

A radical expression is an algebraic construct that involves a root symbol, an index, and a radicand. The general form of a radical expression is
$$\sqrt[n]{a},$$
where n is a positive integer called the index and a is the radicand. In the special case where $n = 2$, the expression is known as the square root and is commonly written as
$$\sqrt{a}.$$

In expressions with an even index, the value of a is usually restricted to nonnegative numbers in order to yield a real number result. The notation indicates that the expression represents the unique nonnegative number which, raised to the power n, equals a.

Fundamental Properties of Radicals

Radical expressions adhere to several key properties that simplify a wide variety of algebraic manipulations. One important property is the product property of radicals. For nonnegative numbers a and

b with a given index n, the property is stated as

$$\sqrt[n]{a \cdot b} = \sqrt[n]{a} \cdot \sqrt[n]{b}.$$

Similarly, the quotient property of radicals asserts that for any nonnegative a and any nonzero b,

$$\sqrt[n]{\frac{a}{b}} = \frac{\sqrt[n]{a}}{\sqrt[n]{b}}.$$

In addition, if a radical expression is raised to a power that is a multiple of the index, the operations can be reversed using the definition of the radical. These properties form the foundation for the manipulation, simplification, and evaluation of radical expressions.

Techniques for Simplifying Radical Expressions

Simplification of a radical expression often begins with identifying and extracting factors that are perfect powers relative to the index. Consider a radical expression such as

$$\sqrt{72}.$$

By expressing 72 as the product of a perfect square and another factor,
$$72 = 36 \times 2,$$
the square root can be rewritten using the product property of radicals:
$$\sqrt{72} = \sqrt{36 \times 2} = \sqrt{36} \cdot \sqrt{2} = 6\sqrt{2}.$$

When simplifying, prime factorization is frequently employed to break down the radicand into its constituent factors. The factors that form a perfect power corresponding to the index can then be extracted outside of the radical sign. Furthermore, combining like radical terms follows the same principle as combining like terms in polynomial expressions, and expressions with identical radicands and indices can be added or subtracted by combining their numerical coefficients.

Operations Involving Radical Expressions

Arithmetic operations with radical expressions require the consistent application of the properties discussed. Multiplication of two radical expressions with the same index can be performed by multiplying the numbers under the radical sign:

$$\sqrt[n]{a} \cdot \sqrt[n]{b} = \sqrt[n]{ab}.$$

Division is handled in a similar manner by employing the quotient property:

$$\frac{\sqrt[n]{a}}{\sqrt[n]{b}} = \sqrt[n]{\frac{a}{b}},$$

providing that $b \neq 0$. In contrast, addition and subtraction of radical expressions necessitate that the radicals be like terms. This involves having the same index and radicand so that expressions such as $3\sqrt{2}$ and $5\sqrt{2}$ may be combined to yield $8\sqrt{2}$. If the radicands differ, the expression remains in its expanded form and the coefficients cannot be directly combined.

Rationalizing the Denominator

In many expressions, the presence of a radical in the denominator is undesirable for the sake of clarity and further manipulation. Rationalizing the denominator is the process by which the radical is eliminated from the denominator. For a simple fraction such as

$$\frac{1}{\sqrt{a}},$$

multiplying both the numerator and the denominator by $\sqrt{a}$ yields

$$\frac{\sqrt{a}}{a}.$$

In more complex expressions, particularly those involving binomials with radical terms, the use of conjugates is an effective method. For example, an expression of the form

$$\frac{1}{\sqrt{a} + \sqrt{b}}$$

can be rationalized by multiplying both the numerator and the denominator by the conjugate $\sqrt{a} - \sqrt{b}$, thereby eliminating the radical terms in the denominator through the difference of squares. The process refines the expression so that further algebraic operations can be carried out with the resulting rational denominator.

Multiple Choice Questions

1. Which of the following represents the general form of a radical expression?

    (a) a^n
    (b) $\sqrt[n]{a}$
    (c) $\sqrt{a^n}$
    (d) $n\sqrt{a}$

2. In radical expressions with an even index, the radicand is typically required to be:

    (a) A positive integer
    (b) A perfect square
    (c) Nonnegative
    (d) Nonzero

3. Which property of radicals is illustrated by the equation
$$\sqrt[n]{a \cdot b} = \sqrt[n]{a} \cdot \sqrt[n]{b}?$$

    (a) Quotient property
    (b) Power property
    (c) Product property
    (d) Distributive property

4. When simplifying the expression $\sqrt{72}$, which factorization is most helpful?

    (a) $72 = 8 \times 9$
    (b) $72 = 36 \times 2$
    (c) $72 = 12 \times 6$
    (d) $72 = 16 \times 4.5$

5. Which statement is true about adding radical expressions?

(a) Radical expressions can be added regardless of their radicands.

(b) Radical expressions with the same index can always be added.

(c) Only radical expressions with identical radicands and indices can be combined.

(d) Radical expressions must be multiplied before being added.

6. What is the primary purpose of rationalizing the denominator in a fraction?

   (a) To simplify the numerator.

   (b) To eliminate radicals from the denominator.

   (c) To combine like radical terms.

   (d) To convert the fraction into a decimal.

7. To rationalize the denominator of
$$\frac{1}{\sqrt{a} + \sqrt{b}},$$
which method is most effective?

   (a) Multiply the numerator and denominator by $\sqrt{a}$.

   (b) Multiply the numerator and denominator by $\sqrt{b}$.

   (c) Multiply the numerator and denominator by $\sqrt{a} - \sqrt{b}$.

   (d) Multiply the numerator and denominator by $\sqrt{a} + \sqrt{b}$.

Answers:

1. **B:** $\sqrt[n]{a}$
   This is the general form of a radical expression where n is the index and a is the radicand.

2. **C: Nonnegative**
   For radicals with an even index, the radicand must be nonnegative to ensure the result is a real number.

3. **C: Product property**
   The equation $\sqrt[n]{a \cdot b} = \sqrt[n]{a} \cdot \sqrt[n]{b}$ exemplifies the product property of radicals.

4. **B: $72 = 36 \times 2$**
   Factoring 72 as 36×2 is effective because 36 is a perfect square ($\sqrt{36} = 6$), which simplifies the expression.

5. **C: Only radical expressions with identical radicands and indices can be combined.**
   Radical expressions may be added or subtracted only if they are like terms, meaning they share the same radicand and index.

6. **B: To eliminate radicals from the denominator.**
   Rationalizing the denominator removes irrational numbers (radicals) from the denominator, making the expression easier to work with.

7. **C: Multiply the numerator and denominator by $\sqrt{a} - \sqrt{b}$.**
   Multiplying by the conjugate $\sqrt{a} - \sqrt{b}$ employs the difference of squares to eliminate the radicals from the denominator.

Practice Problems

1. Simplify the radical expression:
$$\sqrt{72}$$

2. Simplify the product of radicals using the product property:
$$\sqrt{3} \cdot \sqrt{12}$$

3. Simplify the expression by combining like radical terms:
$$2\sqrt{18} + 3\sqrt{8} - \sqrt{50}$$

4. Rationalize the denominator of the following expression:
$$\frac{1}{\sqrt{5} + \sqrt{3}}$$

5. Solve the radical equation for x:
$$\sqrt{x+5} = 3$$

6. Simplify the cube root expression:
$$\sqrt[3]{54}$$

Answers

1. **Solution:** To simplify
   $$\sqrt{72},$$
   we first factor 72 into a product where one factor is a perfect square. Notice that
   $$72 = 36 \times 2,$$
   and since 36 is a perfect square, we can write:
   $$\sqrt{72} = \sqrt{36 \times 2} = \sqrt{36} \cdot \sqrt{2} = 6\sqrt{2}.$$
   Therefore,
   $$\sqrt{72} = 6\sqrt{2}.$$

2. **Solution:** Using the product property of radicals, we know that:
   $$\sqrt{3} \cdot \sqrt{12} = \sqrt{3 \times 12} = \sqrt{36}.$$
   Since 36 is a perfect square,
   $$\sqrt{36} = 6.$$
   Thus, the simplified product is
   $$6.$$

3. **Solution:** We simplify each radical term by extracting perfect square factors:

   First, for $\sqrt{18}$:
   $$\sqrt{18} = \sqrt{9 \times 2} = 3\sqrt{2},$$
   so
   $$2\sqrt{18} = 2 \times 3\sqrt{2} = 6\sqrt{2}.$$

   Next, for $\sqrt{8}$:
   $$\sqrt{8} = \sqrt{4 \times 2} = 2\sqrt{2},$$
   hence,
   $$3\sqrt{8} = 3 \times 2\sqrt{2} = 6\sqrt{2}.$$

   Finally, for $\sqrt{50}$:
   $$\sqrt{50} = \sqrt{25 \times 2} = 5\sqrt{2}.$$

   Now, combine the like terms:
   $$6\sqrt{2} + 6\sqrt{2} - 5\sqrt{2} = (6 + 6 - 5)\sqrt{2} = 7\sqrt{2}.$$

   Therefore,
   $$2\sqrt{18} + 3\sqrt{8} - \sqrt{50} = 7\sqrt{2}.$$

4. **Solution:** To rationalize the denominator of
   $$\frac{1}{\sqrt{5} + \sqrt{3}},$$
   we multiply the numerator and the denominator by the conjugate of the denominator, which is $\sqrt{5} - \sqrt{3}$:
   $$\frac{1}{\sqrt{5} + \sqrt{3}} \times \frac{\sqrt{5} - \sqrt{3}}{\sqrt{5} - \sqrt{3}} = \frac{\sqrt{5} - \sqrt{3}}{(\sqrt{5} + \sqrt{3})(\sqrt{5} - \sqrt{3})}.$$

   The denominator simplifies by the difference of squares:
   $$(\sqrt{5} + \sqrt{3})(\sqrt{5} - \sqrt{3}) = (\sqrt{5})^2 - (\sqrt{3})^2 = 5 - 3 = 2.$$

   Thus, the rationalized form is
   $$\frac{\sqrt{5} - \sqrt{3}}{2}.$$

5. **Solution:** To solve the equation
$$\sqrt{x+5} = 3,$$
we first eliminate the square root by squaring both sides:
$$(\sqrt{x+5})^2 = 3^2,$$
which gives:
$$x + 5 = 9.$$
Subtract 5 from both sides:
$$x = 9 - 5 = 4.$$
To verify, substitute $x = 4$ back into the original equation:
$$\sqrt{4+5} = \sqrt{9} = 3.$$
The solution checks out, so
$$x = 4.$$

6. **Solution:** To simplify the cube root
$$\sqrt[3]{54},$$
we express 54 as the product of a perfect cube and another factor. Notice that
$$54 = 27 \times 2,$$
and 27 is a perfect cube because
$$27 = 3^3.$$
Thus, we have:
$$\sqrt[3]{54} = \sqrt[3]{27 \times 2} = \sqrt[3]{27} \cdot \sqrt[3]{2} = 3\sqrt[3]{2}.$$
Therefore,
$$\sqrt[3]{54} = 3\sqrt[3]{2}.$$

Chapter 30

Operations and Equations Involving Radicals

Basic Concepts and Simplification of Radical Expressions

A radical expression is denoted by the general form

$$\sqrt[n]{a},$$

where the positive integer n is referred to as the index and a is the radicand. In many cases, the process of simplification involves factoring the radicand to identify any perfect nth powers contained within it. When such factors exist, they may be extracted from under the radical sign to yield a simplified expression. For instance, if the radicand is expressed as a product of a perfect square and another factor, then the square root may be rewritten as the product of the square roots of these factors. The basic properties governing radical expressions, such as the product property

$$\sqrt[n]{a} \cdot \sqrt[n]{b} = \sqrt[n]{ab},$$

and the quotient property

$$\sqrt[n]{\frac{a}{b}} = \frac{\sqrt[n]{a}}{\sqrt[n]{b}},$$

are fundamental to the systematic approach aimed at reducing and rewriting radical expressions in their simplest form.

Arithmetic Operations on Radical Expressions

1 Addition and Subtraction of Radical Expressions

Addition and subtraction involving radical expressions require that the radicals be like terms, meaning the expressions must share identical indices and radicands. When two radical expressions are like terms, their coefficients are combined while the radical part remains unchanged. For example, the expressions

$$3\sqrt{2} \quad \text{and} \quad 5\sqrt{2}$$

can be combined to yield

$$(3+5)\sqrt{2} = 8\sqrt{2}.$$

In contrast, when the radicands or the indices differ, the expressions are not combinable in this direct manner and must be left as separate terms unless further manipulation makes them alike.

2 Multiplication of Radical Expressions

Multiplication of radical expressions utilizes the product property of radicals. When multiplying two radicals with the same index, the radicands are multiplied together, and the radical encompassing the product encapsulates the result:

$$\sqrt[n]{a} \cdot \sqrt[n]{b} = \sqrt[n]{ab}.$$

For instance, the multiplication of the radicals

$$\sqrt{3} \quad \text{and} \quad \sqrt{12}$$

results in

$$\sqrt{3} \cdot \sqrt{12} = \sqrt{36},$$

which simplifies further to produce the value 6. Multiplicative operations often require post-processing to extract perfect powers from the resulting radicand, thereby achieving a more refined expression.

Solving Equations Involving Radical Expressions

Equations that contain radical expressions require a methodical approach to isolate and subsequently eliminate the radical component. The standard procedure begins with isolating the radical expression on one side of the equation, a critical step that prepares the equation for the elimination of the radical by raising both sides to the power corresponding to the index of the radical. Consider an equation of the form

$$\sqrt{x + c} = d,$$

where c and d are constants. By isolating the radical term, squaring both sides leads to

$$x + c = d^2.$$

The resulting equation is then solved using conventional algebraic techniques.

When an equation contains more than one radical term, it becomes necessary to isolate one radical at a time. Each radical is eliminated through successive exponentiation steps, bearing in mind that these operations may introduce extraneous solutions. A careful verification process is essential after the algebraic solution has been determined in order to confirm that each candidate solution satisfies the original equation and adheres to the domain restrictions imposed by the presence of the radical.

The strategy for solving radical equations involves three primary steps:

1. Isolation of the radical expression.

2. Application of the appropriate exponent to both sides in order to remove the radical.

3. Resolution of the resulting equation, followed by substitution of each solution back into the original equation to verify correctness.

This thorough procedure ensures that all valid solutions are identified while any extraneous solutions are rejected.

Multiple Choice Questions

1. Which of the following represents the general form of a radical expression?

    (a) $\sqrt{n}a$
    (b) $a^{\frac{1}{n}}$
    (c) $\sqrt[n]{a}$
    (d) $\sqrt{an}$

2. Which property allows you to multiply two radical expressions with the same index?

    (a) Quotient Property
    (b) Product Property
    (c) Distributive Property
    (d) Commutative Property

3. In simplifying the radical expression $\sqrt{72}$, which factorization leads to the simplest form?

    (a) $72 = 8 \times 9$, so $\sqrt{72} = \sqrt{8}\sqrt{9}$
    (b) $72 = 36 \times 2$, so $\sqrt{72} = \sqrt{36}\sqrt{2} = 6\sqrt{2}$
    (c) $72 = 12 \times 6$, so $\sqrt{72} = \sqrt{12}\sqrt{6}$
    (d) $72 = 18 \times 4$, so $\sqrt{72} = \sqrt{18}\sqrt{4}$

4. Addition and subtraction of radical expressions can be performed directly only when:

    (a) They have the same coefficient.
    (b) They are both in simplest radical form.
    (c) They are like terms—that is, they have the same index and the same radicand.
    (d) They are square roots.

5. What is the correct first step in solving the equation $\sqrt{x+4} = 7$?

    (a) Isolate the radical term, then square both sides.
    (b) Square both sides immediately.
    (c) Add 4 to both sides.

(d) Multiply both sides by 7.

6. After squaring both sides of a radical equation, why is it necessary to check each solution by substituting back into the original equation?

   (a) To simplify the final answer further.
   (b) To ensure no extraneous solutions were introduced.
   (c) To eliminate negative numbers from the solution.
   (d) To verify that all radicals have been completely removed.

7. Consider the equation $\sqrt{3x-5} = 4$. Which of the following is the correct solution for x?

   (a) $x = \frac{7}{3}$
   (b) $x = 7$
   (c) $x = \frac{16}{3}$
   (d) $x = 21$

Answers:

1. **C:** $\sqrt[n]{a}$
   This is the general form of a radical expression where n is the index (a positive integer) and a is the radicand.

2. **B: Product Property**
   The product property of radicals states that $\sqrt[n]{a} \cdot \sqrt[n]{b} = \sqrt[n]{ab}$ when the radicals have the same index, making multiplication of such expressions straightforward.

3. **B:** $72 = 36 \times 2$, so $\sqrt{72} = 6\sqrt{2}$
   Factoring 72 as 36×2 allows us to extract the perfect square 36 from under the radical, simplifying the expression to $6\sqrt{2}$.

4. **C: They are like terms—that is, they have the same index and the same radicand.**
   Radical expressions can only be added or subtracted when they are like terms, meaning both the radicand and the index must match.

5. **A: Isolate the radical term, then square both sides.**
   The proper method for solving a radical equation begins with isolating the radical term, which then allows you to eliminate the radical by squaring both sides of the equation.

6. **B: To ensure no extraneous solutions were introduced.**
   Squaring both sides of an equation can introduce extraneous solutions that do not satisfy the original equation; therefore, it is critical to check each potential solution by substituting it back into the original equation.

7. **B:** $x = 7$
   Squaring both sides of $\sqrt{3x - 5} = 4$ gives $3x - 5 = 16$. Solving for x leads to $3x = 21$, so $x = 7$. This solution must then be verified in the original equation to ensure it is valid.

Practice Problems

1. Simplify the following radical expression:
   $$\sqrt{72}$$

2. Simplify and combine the following radical expressions:
   $$3\sqrt{8} + 2\sqrt{18}$$

3. Multiply the following radical expressions and simplify:
$$\sqrt{3} \cdot \sqrt{12}$$

4. Simplify the quotient of the following radicals:
$$\frac{\sqrt{50}}{\sqrt{2}}$$

5. Solve the following radical equation:
$$\sqrt{x+4} = 6$$

6. Solve for x in the following radical equation and check for extraneous solutions:
$$\sqrt{2x-1} = x - 3$$

Answers

1. **Simplify $\sqrt{72}$.**

    Solution:

    Factor 72 as
    $$72 = 36 \times 2.$$
    Then, using the property of radicals,
    $$\sqrt{72} = \sqrt{36 \times 2} = \sqrt{36}\sqrt{2} = 6\sqrt{2}.$$

    Explanation: Identify the perfect square 36 within the radicand and apply the product property of radicals to simplify the expression.

2. **Simplify and combine $3\sqrt{8} + 2\sqrt{18}$.**

    Solution:

    First, simplify each radical:
    $$\sqrt{8} = \sqrt{4 \times 2} = 2\sqrt{2}, \quad \sqrt{18} = \sqrt{9 \times 2} = 3\sqrt{2}.$$
    Then,
    $$3\sqrt{8} = 3 \cdot 2\sqrt{2} = 6\sqrt{2}, \quad 2\sqrt{18} = 2 \cdot 3\sqrt{2} = 6\sqrt{2}.$$

Combining these like terms yields
$$6\sqrt{2} + 6\sqrt{2} = 12\sqrt{2}.$$

Explanation: Both radical terms simplify to expressions involving $\sqrt{2}$. Once expressed as like terms, add their coefficients to obtain the simplified result.

3. **Multiply** $\sqrt{3} \cdot \sqrt{12}$.

   Solution:

   By using the product property of radicals,
   $$\sqrt{3} \cdot \sqrt{12} = \sqrt{3 \times 12} = \sqrt{36} = 6.$$

   Explanation: Multiplying the radicands results in a perfect square, making the simplification straightforward.

4. **Simplify** $\frac{\sqrt{50}}{\sqrt{2}}$.

   Solution:

   Using the quotient property of radicals,
   $$\frac{\sqrt{50}}{\sqrt{2}} = \sqrt{\frac{50}{2}} = \sqrt{25} = 5.$$

   Explanation: Division under the radical sign simplifies the radicand to 25, which is a perfect square, yielding the simplified value 5.

5. **Solve** the equation $\sqrt{x+4} = 6$.

   Solution:

   Square both sides to eliminate the radical:
   $$(\sqrt{x+4})^2 = 6^2,$$
   $$x + 4 = 36.$$

   Subtract 4 from both sides:
   $$x = 32.$$

   Explanation: Isolating the radical and squaring both sides converts the radical equation into a linear equation. The solution $x = 32$ satisfies the condition that the expression under the square root remains nonnegative.

6. **Solve** the equation $\sqrt{2x-1} = x-3$ and check for extraneous solutions.

   Solution:

   Step 1: Domain Restrictions

   The expression under the square root must be nonnegative:
   $$2x - 1 \geq 0 \quad \Rightarrow \quad x \geq \frac{1}{2}.$$

   Also, since the square root produces a nonnegative result, the right side must satisfy:
   $$x - 3 \geq 0 \quad \Rightarrow \quad x \geq 3.$$

   Therefore, we require $x \geq 3$.

   Step 2: Square Both Sides

   Square the equation:
   $$(\sqrt{2x-1})^2 = (x-3)^2,$$
   $$2x - 1 = x^2 - 6x + 9.$$

   Step 3: Rearrange to Form a Quadratic Equation

   Bring all terms to one side:
   $$0 = x^2 - 6x + 9 - 2x + 1,$$
   $$0 = x^2 - 8x + 10.$$

   Step 4: Solve the Quadratic Equation

   Use the quadratic formula:
   $$x = \frac{8 \pm \sqrt{64-40}}{2} = \frac{8 \pm \sqrt{24}}{2} = \frac{8 \pm 2\sqrt{6}}{2} = 4 \pm \sqrt{6}.$$

   Step 5: Check Against Domain Restrictions

   For $x = 4 + \sqrt{6}$ (approximately 6.45), the condition $x \geq 3$ is satisfied.

   For $x = 4 - \sqrt{6}$ (approximately 1.55), the condition $x \geq 3$ is not met, and substituting it into the original equation yields a negative value on the right side, which is not possible.

Thus, the only valid solution is:
$$x = 4 + \sqrt{6}.$$

Explanation: After squaring both sides, the equation becomes quadratic and yields two solutions. Domain restrictions for the square root and the structure of the original equation eliminate $x = 4 - \sqrt{6}$ as an extraneous solution. Only $x = 4 + \sqrt{6}$ satisfies all necessary conditions.

Chapter 31

Quadratic Equations: Forms and Strategies

Forms of Quadratic Equations

A quadratic equation is defined by the presence of a variable raised to the power of two. The standard form of a quadratic equation is written as
$$ax^2 + bx + c = 0,$$
where a, b, and c are constants with $a \neq 0$. This form is advantageous for applying various solution strategies and for a direct interpretation of the coefficients. Alternatively, the same quadratic expression may be written in vertex form as
$$a(x-h)^2 + k = 0,$$
where the coordinates (h, k) represent the vertex of the associated parabola, providing insight into the graph's symmetry and its extreme value. When the quadratic expression is factorable, it can also be presented in factored form:
$$a(x - r_1)(x - r_2) = 0,$$
with r_1 and r_2 representing the roots or zeros of the quadratic function. These equivalent forms allow for different analytical perspectives and suggest distinct solution paths.

Strategies for Solving Quadratic Equations

Various algebraic techniques can be employed to solve quadratic equations. The method chosen is often influenced by the specific form of the equation and the characteristics of its coefficients.

1 Factoring Methods

When the quadratic expression is factorable, it is rewritten as
$$a(x - r_1)(x - r_2) = 0.$$
The zero-product property is then applied, leading to the equations
$$x - r_1 = 0 \quad \text{or} \quad x - r_2 = 0.$$
This approach yields the solutions $x = r_1$ and $x = r_2$. The factoring process involves identifying two numbers whose product is equal to ac and whose sum is equal to b, considering the original form $ax^2 + bx + c$. This method is elegant when the factors are integers or simple fractions, offering a direct route to the solutions.

2 Completing the Square

Completing the square systematically converts a quadratic equation in standard form into a perfect square. Beginning with
$$ax^2 + bx + c = 0,$$
the equation is first normalized by dividing all terms by a, resulting in
$$x^2 + \frac{b}{a}x + \frac{c}{a} = 0.$$
To form a perfect square trinomial, the term $\left(\frac{b}{2a}\right)^2$ is added and subtracted. This permits the equation to be rewritten as
$$\left(x + \frac{b}{2a}\right)^2 - \left(\frac{b}{2a}\right)^2 + \frac{c}{a} = 0.$$
Isolating the perfect square yields
$$\left(x + \frac{b}{2a}\right)^2 = \left(\frac{b}{2a}\right)^2 - \frac{c}{a}.$$
Taking the square root of both sides and then solving for x produces the solutions. This method not only unlocks the value of x but also reveals the vertex of the corresponding quadratic function through the term $x + \frac{b}{2a}$.

3 Quadratic Formula

The quadratic formula serves as a comprehensive tool for solving any quadratic equation expressed in standard form. Derived through the process of completing the square, the formula is stated as
$$x = \frac{-b \pm \sqrt{b^2 - 4ac}}{2a}.$$
Within this formula, the expression $b^2 - 4ac$ is known as the discriminant. Its value determines the nature of the roots: when the discriminant is positive, the quadratic has two distinct real solutions; when it is zero, the equation has one repeated real solution; and when it is negative, the solutions are complex conjugates. This formula encapsulates all potential cases and provides a uniform procedure for obtaining the roots regardless of the factorability of the quadratic expression.

4 Analysis of the Discriminant

The discriminant, given by
$$b^2 - 4ac,$$
offers critical information regarding the types of solutions that may be expected. A positive value for the discriminant implies that the quadratic equation has two distinct real roots, corresponding to two intersecting points with the x-axis. A zero value indicates that the vertex of the parabola lies exactly on the x-axis, and hence the equation has a single (repeated) real solution. A negative discriminant reveals that the quadratic equation has no real solutions; instead, the roots are complex numbers. Through careful evaluation of the discriminant, the behavior of the quadratic function can be discerned prior to a complete algebraic resolution of the equation.

Applications of Algebraic Methods

The various methods for solving quadratic equations are interconnected and can often be used to reinforce each other. Conversion from the standard form to the vertex form through completing the square elucidates the symmetry of the parabola and identifies the vertex explicitly. Factoring, when it is applicable, provides a quick and intuitive method for reaching the solutions via the zero-product property. The quadratic formula stands as the most general method, capable of addressing all quadratic equations regardless of the simplicity of their factors.

These algebraic techniques enhance comprehension of the underlying structure of quadratic equations and bolster the ability to analyze their behavior both algebraically and graphically. The detailed examination of different forms and solution strategies affords a solid foundation in the study of quadratic functions.

Multiple Choice Questions

1. Which of the following equations is written in the standard form of a quadratic equation?

    (a) $ax^2 + bx + c = 0$, where a, b, and c are constants with a 0

    (b) $a(x - h)^2 + k = 0$, where (h, k) is the vertex of the parabola

    (c) $a(x - r)(x - r) = 0$, where r and r are the roots of the equation

    (d) None of the above

2. Which form of a quadratic equation explicitly reveals the vertex of its parabola?

    (a) Standard form: $ax^2 + bx + c = 0$

    (b) Vertex form: $a(x - h)^2 + k = 0$

    (c) Factored form: $a(x - r)(x - r) = 0$

    (d) All of the above

3. When a quadratic expression is easily factorable, what is the simplest method to solve the equation?

    (a) Completing the square

    (b) Factoring

    (c) Using the quadratic formula

    (d) Graphing the equation

4. In the process of completing the square for the quadratic equation $ax^2 + bx + c = 0$, what is the first critical step?

(a) Factor out the constant term c

(b) Divide the entire equation by a to normalize the coefficient of x^2

(c) Subtract c from both sides of the equation

(d) Add (b/2a) to both sides

5. The discriminant in the quadratic formula is given by $b^2 - 4ac$. What information does it provide about the quadratic equation?

    (a) It gives the sum of the roots

    (b) It determines the number and type (real or complex) of the roots

    (c) It identifies the axis of symmetry of the parabola

    (d) It calculates the vertex coordinates of the quadratic function

6. Which of the following is the correct quadratic formula?

    (a) $x = \dfrac{-b \pm \sqrt{b^2 - 4ac}}{2a}$

    (b) $x = \dfrac{b \pm \sqrt{b^2 - 4ac}}{2a}$

    (c) $x = \dfrac{-b \pm \sqrt{4ac - b^2}}{2a}$

    (d) $x = \dfrac{b \pm \sqrt{4ac - b^2}}{2a}$

7. Converting a quadratic equation from standard form to vertex form by completing the square is particularly useful because it:

    (a) Simplifies the factoring process

    (b) Reveals the axis of symmetry and the vertex of the parabola

    (c) Always yields real roots

    (d) Eliminates the need to use the quadratic formula

Answers:

1. **A:** $ax^2 + bx + c = 0$, where a, b, and c are constants with a 0. This is the standard form of a quadratic equation, which is essential for applying many solution strategies.

2. **B:** Vertex form: $a(x - h)^2 + k = 0$. The vertex form directly displays the vertex (h, k) of the parabola, making it easier to graph and analyze its symmetry.

3. **B:** Factoring. When a quadratic is easily factorable, rewriting it as $a(x - r)(x - r) = 0$ allows the use of the zero-product property to quickly find the roots.

4. **B:** Divide the entire equation by a to normalize the coefficient of x^2. Normalization (making the coefficient of x^2 equal to 1) is the crucial first step in completing the square, enabling the formation of a perfect square trinomial.

5. **B:** It determines the number and type (real or complex) of the roots. The discriminant, $b^2 - 4ac$, indicates whether the quadratic has two distinct real roots (if positive), one repeated real root (if zero), or two complex roots (if negative).

6. **A:** $x = \dfrac{-b \pm \sqrt{b^2 - 4ac}}{2a}$. This formula, derived from completing the square, provides a universal method for solving any quadratic equation regardless of its factorability.

7. **B:** Reveals the axis of symmetry and the vertex of the parabola. Converting to vertex form by completing the square not only facilitates solving the equation but also makes the key graphical features—the vertex and axis of symmetry—readily apparent.

Practice Problems

1. Convert the quadratic equation

$$2x^2 - 8x + 6 = 0$$

   into vertex form by completing the square, and state the vertex of the parabola.

2. Factor the quadratic equation
$$6x^2 - 11x + 3 = 0$$
by factoring, and find the solutions for x.

3. Solve the quadratic equation
$$x^2 + 4x + 5 = 0$$
using the quadratic formula, and analyze the discriminant to determine the nature of the roots.

4. Determine why the quadratic equation
$$4x^2 - 12x + 9 = 0$$
has a repeated real root, and find that root.

5. Rewrite the quadratic equation
$$3x^2 - 6x + 2 = 0$$
in vertex form by completing the square, and identify the vertex of the parabola.

6. Use the quadratic formula to solve the equation
$$2x^2 + 3x - 5 = 0$$
and verify your solution by analyzing the discriminant.

Answers

1. **Solution:** We start with the quadratic equation
$$2x^2 - 8x + 6 = 0.$$
First, factor out the coefficient of x^2:
$$2\left(x^2 - 4x\right) + 6 = 0.$$
Next, complete the square for the expression in parentheses. The term needed is
$$\left(\frac{-4}{2}\right)^2 = 4.$$

Add and subtract 4 inside the parentheses:
$$2\left[(x^2 - 4x + 4) - 4\right] + 6 = 0.$$

Recognize the perfect square:
$$2\left[(x - 2)^2 - 4\right] + 6 = 0.$$

Distribute the 2:
$$2(x - 2)^2 - 8 + 6 = 0.$$

Simplify the constants:
$$2(x - 2)^2 - 2 = 0.$$

This is the vertex form, and the vertex of the parabola is the point $(2, -2)$.

2. **Solution:** Consider the quadratic equation
$$6x^2 - 11x + 3 = 0.$$

To factor, find two numbers that multiply to $6 \times 3 = 18$ and add to -11. These numbers are -9 and -2. Rewrite the middle term:
$$6x^2 - 9x - 2x + 3 = 0.$$

Factor by grouping:
$$3x(2x - 3) - 1(2x - 3) = 0.$$

Factor out the common binomial:
$$(2x - 3)(3x - 1) = 0.$$

Setting each factor equal to zero:
$$2x - 3 = 0 \quad \text{or} \quad 3x - 1 = 0.$$

Solving these gives:
$$x = \frac{3}{2} \quad \text{or} \quad x = \frac{1}{3}.$$

3. **Solution:** To solve
$$x^2 + 4x + 5 = 0,$$
we apply the quadratic formula:
$$x = \frac{-b \pm \sqrt{b^2 - 4ac}}{2a},$$
where $a = 1$, $b = 4$, and $c = 5$. First, compute the discriminant:
$$b^2 - 4ac = 4^2 - 4(1)(5) = 16 - 20 = -4.$$
Since the discriminant is negative, the equation has two complex solutions. Substituting into the formula:
$$x = \frac{-4 \pm \sqrt{-4}}{2} = \frac{-4 \pm 2i}{2} = -2 \pm i.$$
The negative discriminant confirms there are no real roots, only the complex conjugate pair $x = -2 + i$ and $x = -2 - i$.

4. **Solution:** The quadratic equation
$$4x^2 - 12x + 9 = 0$$
can be recognized as a perfect square because
$$4x^2 - 12x + 9 = (2x - 3)^2.$$
Setting the perfect square equal to zero:
$$(2x - 3)^2 = 0 \quad \Rightarrow \quad 2x - 3 = 0.$$
Solving for x, we find:
$$x = \frac{3}{2}.$$
Additionally, the discriminant calculated as
$$(-12)^2 - 4(4)(9) = 144 - 144 = 0,$$
confirms that there is exactly one repeated (double) real root, which is $x = \frac{3}{2}$.

5. **Solution:** Consider the quadratic equation
$$3x^2 - 6x + 2 = 0.$$
To write it in vertex form, first factor out the coefficient of x^2:
$$3\left(x^2 - 2x\right) + 2 = 0.$$
Complete the square for the expression $x^2 - 2x$ by adding and subtracting $\left(\frac{-2}{2}\right)^2 = 1$:
$$3\left[\left(x^2 - 2x + 1\right) - 1\right] + 2 = 0.$$
Rewrite the perfect square:
$$3\left[(x-1)^2 - 1\right] + 2 = 0.$$
Distribute the 3:
$$3(x-1)^2 - 3 + 2 = 0.$$
Simplify the constants:
$$3(x-1)^2 - 1 = 0.$$
Therefore, the vertex form is
$$3(x-1)^2 - 1 = 0,$$
and the vertex of the parabola is $(1, -1)$.

6. **Solution:** To solve
$$2x^2 + 3x - 5 = 0,$$
we use the quadratic formula with $a = 2$, $b = 3$, and $c = -5$:
$$x = \frac{-b \pm \sqrt{b^2 - 4ac}}{2a}.$$
First, compute the discriminant:
$$b^2 - 4ac = 3^2 - 4(2)(-5) = 9 + 40 = 49.$$
Since the discriminant is positive and a perfect square, there are two distinct real solutions. Substitute the values into the formula:
$$x = \frac{-3 \pm \sqrt{49}}{2 \cdot 2} = \frac{-3 \pm 7}{4}.$$

349

This gives:
$$x = \frac{-3+7}{4} = \frac{4}{4} = 1,$$
and
$$x = \frac{-3-7}{4} = \frac{-10}{4} = -\frac{5}{2}.$$

The discriminant 49 confirms that the roots are real and distinct, with the solutions being $x = 1$ and $x = -\frac{5}{2}$.

Chapter 32

Solving Quadratic Equations by Factoring

Quadratic Equations in Standard Form

A quadratic equation is typically expressed in the form

$$ax^2 + bx + c = 0,$$

where a, b, and c are constants with $a \neq 0$. This format serves as the foundation for the factoring method. When the quadratic expression is arranged in this form, the procedure for decomposing it into a product of simpler expressions is well defined. In instances where the coefficient a exceeds one, factoring out any common factors at the outset simplifies subsequent steps.

Methods for Factoring Quadratic Expressions

Factoring involves rewriting a quadratic expression as the product of two binomial expressions. When a quadratic is factorable, it assumes the structure

$$a(x - r_1)(x - r_2) = 0,$$

with r_1 and r_2 representing the zeros of the equation. A variety of techniques assist in this process and include the extraction of common factors, the splitting of the middle term, and the recognition of special algebraic patterns.

1 Extracting Common Factors

The first analytical step is to determine whether a common factor exists among all the terms. The extraction of the greatest common factor reduces the quadratic to a simpler form, thus making the factorization process more transparent. For example, consider the quadratic expression
$$2x^2 + 4x - 6.$$
A common factor of 2 can be factored out:
$$2\left(x^2 + 2x - 3\right).$$
The expression within the parentheses is then examined and, if possible, factored further.

2 Splitting the Middle Term

A systematic method for factoring quadratics involves splitting the middle term. In a quadratic given by
$$ax^2 + bx + c = 0,$$
two numbers are sought that multiply to ac and sum to b. Once these two numbers are identified, the middle term bx is expressed as the sum of two terms corresponding to these numbers. For instance, in the quadratic
$$x^2 + 5x + 6,$$
the numbers 2 and 3 satisfy the requirements since $2 \times 3 = 6$ and $2 + 3 = 5$. Rewriting the expression yields
$$x^2 + 2x + 3x + 6,$$
which can be grouped as
$$x(x + 2) + 3(x + 2).$$
The common binomial factor $(x+2)$ is then factored out, resulting in
$$(x + 3)(x + 2).$$

3 Recognition of Special Patterns

Certain quadratic expressions conform to recognizable patterns, such as perfect square trinomials or the difference of squares. A perfect square trinomial appears as

$$(x + d)^2 = x^2 + 2dx + d^2,$$

while the difference of squares is expressed by

$$x^2 - d^2 = (x - d)(x + d).$$

Identifying such patterns allows for an immediate factorization into simpler binomial forms, thereby streamlining the solving process.

Application of the Zero Product Property

Once the quadratic equation has been factored into the form

$$a(x - r_1)(x - r_2) = 0,$$

the zero product property is applied to determine the solutions. This property asserts that if a product of factors is equal to zero, then at least one of the factors must itself be zero. Accordingly, the equation

$$(x - r_1)(x - r_2) = 0$$

yields two separate linear equations:

$$x - r_1 = 0 \quad \text{or} \quad x - r_2 = 0.$$

Solving these equations provides the roots $x = r_1$ and $x = r_2$.

Worked Examples

1 Example 1

Consider the quadratic equation

$$x^2 + 5x + 6 = 0.$$

The equation factors neatly as

$$(x + 2)(x + 3) = 0.$$

Application of the zero product property leads to the equations
$$x+2=0 \quad \text{or} \quad x+3=0,$$
so that the solutions are
$$x=-2 \quad \text{and} \quad x=-3.$$

2 Example 2

Examine the quadratic equation
$$2x^2 - x - 6 = 0.$$
First, calculate the product ac where $a=2$ and $c=-6$:
$$ac = -12.$$
Determine two numbers that multiply to -12 and add to $b=-1$. The numbers -4 and 3 fulfill these conditions. Rewrite the equation as
$$2x^2 - 4x + 3x - 6 = 0.$$
Grouping terms accordingly:
$$(2x^2 - 4x) + (3x - 6) = 0,$$
factorization within each group gives:
$$2x(x-2) + 3(x-2) = 0.$$
The common factor $(x-2)$ is extracted to yield:
$$(2x+3)(x-2) = 0.$$
The zero product property then provides:
$$2x+3=0 \quad \text{or} \quad x-2=0,$$
resulting in the solutions
$$x = -\frac{3}{2} \quad \text{and} \quad x=2.$$

3 Example 3

Consider the quadratic equation that exhibits a common factor:

$$4x^2 + 8x - 12 = 0.$$

Factoring out the greatest common factor of 4 produces:

$$4(x^2 + 2x - 3) = 0.$$

The quadratic within the parentheses factors as

$$(x + 3)(x - 1),$$

so that the equation becomes:

$$4(x + 3)(x - 1) = 0.$$

The zero product property then implies:

$$x + 3 = 0 \quad \text{or} \quad x - 1 = 0,$$

yielding the solutions

$$x = -3 \quad \text{and} \quad x = 1.$$

Multiple Choice Questions

1. Which of the following represents the standard form of a quadratic equation used for factoring?

   (a) $ax^2 + bx + c = 0$
   (b) $ax + bx^2 + c = 0$
   (c) $x^2 + bx = c$
   (d) $ax^2 - bx + c = 0$

2. Which method for factoring a quadratic equation involves finding two numbers that multiply to ac and add to b?

   (a) Using the quadratic formula
   (b) Splitting the middle term
   (c) Completing the square
   (d) Extracting the greatest common factor

3. Why is it important to factor out the greatest common factor (GCF) before attempting to further factor a quadratic expression?

   (a) It simplifies the equation, making factoring easier.
   (b) It changes the sign of the quadratic expression.
   (c) It always results in a perfect square trinomial.
   (d) It eliminates the need to use the zero product property.

4. After factoring a quadratic equation to the form $a(x-r_1)(x-r_2) = 0$, which property allows you to set each binomial factor equal to zero to solve for x?

   (a) Distributive property
   (b) Commutative property
   (c) Zero product property
   (d) Associative property

5. Which of the following quadratic expressions is a perfect square trinomial?

   (a) $x^2 + 6x + 9$
   (b) $x^2 + 5x + 6$
   (c) $x^2 - 9$
   (d) $2x^2 + 8x + 8$

6. Consider the quadratic equation
$$2x^2 - x - 6 = 0.$$
Which set of factors correctly represents its factorized form?

   (a) $(2x - 3)(x + 2)$
   (b) $(2x + 3)(x - 2)$
   (c) $(x + 3)(2x - 2)$
   (d) $(2x - 3)(x - 2)$

7. Which technique involves rewriting the middle term of a quadratic expression as two terms that can then be grouped for factorization?

   (a) Completing the square

(b) Extracting common factors

(c) Splitting the middle term

(d) Synthetic division

Answers:

1. **A:** $ax^2 + bx + c = 0$
   Explanation: This is the standard form used for quadratic equations, ensuring that the quadratic term is first and that the constant term is last, with $a \neq 0$.

2. **B:** Splitting the middle term
   Explanation: Splitting the middle term involves finding two numbers that multiply to ac and add to b, then rewriting the quadratic accordingly so that it can be factored by grouping.

3. **A:** It simplifies the equation, making factoring easier.
   Explanation: Factoring out the greatest common factor (GCF) reduces the coefficients and overall complexity of the expression, which can make the subsequent factoring steps more manageable.

4. **C:** Zero product property
   Explanation: The zero product property states that if a product of factors is zero, then at least one of the factors must be zero. This allows us to set each factor in the equation $a(x - r_1)(x - r_2) = 0$ equal to zero and solve for x.

5. **A:** $x^2 + 6x + 9$
   Explanation: The quadratic $x^2 + 6x + 9$ is a perfect square trinomial because it can be written as $(x+3)^2$, which fits the pattern $(x + d)^2 = x^2 + 2dx + d^2$.

6. **B:** $(2x + 3)(x - 2)$
   Explanation: Factoring $2x^2 - x - 6 = 0$ by splitting the middle term yields $(2x + 3)(x - 2) = 0$, since expanding these factors reproduces the original quadratic equation.

7. **C:** Splitting the middle term
   Explanation: This technique involves rewriting the middle term into two terms that allow the quadratic expression to be grouped into pairs, each of which can be factored. It is a vital step in the factoring process when the quadratic does not factor easily at first glance.

Practice Problems

1. Factor and solve the quadratic equation:
$$x^2 + 5x + 6 = 0.$$

2. Factor and solve the quadratic equation:
$$2x^2 - x - 6 = 0.$$

3. Factor and solve the quadratic equation:
$$4x^2 + 8x - 12 = 0.$$

4. Solve by factoring:
$$3x^2 - 7x + 2 = 0.$$

5. Solve by factoring:
$$6x^2 + 11x - 35 = 0.$$

6. Factor and solve the quadratic equation (perfect square trinomial):
$$x^2 - 6x + 9 = 0.$$

Answers

1. For the equation
$$x^2 + 5x + 6 = 0,$$
we look for two numbers whose product is 6 and whose sum is 5. The numbers 2 and 3 satisfy these conditions because
$$2 \times 3 = 6 \quad \text{and} \quad 2 + 3 = 5.$$
Thus, we can factor the equation as
$$(x + 2)(x + 3) = 0.$$
By the zero product property, if
$$(x + 2)(x + 3) = 0,$$
then either
$$x + 2 = 0 \quad \text{or} \quad x + 3 = 0.$$
This gives the solutions
$$x = -2 \quad \text{and} \quad x = -3.$$

2. For the equation
$$2x^2 - x - 6 = 0,$$
we first calculate the product $a \times c$ where $a = 2$ and $c = -6$:
$$2 \times (-6) = -12.$$
We search for two numbers that multiply to -12 and add to the coefficient $b = -1$. The numbers -4 and 3 work since
$$(-4) \times 3 = -12 \quad \text{and} \quad (-4) + 3 = -1.$$
We then split the middle term:
$$2x^2 - 4x + 3x - 6 = 0.$$
Grouping the terms, we have:
$$(2x^2 - 4x) + (3x - 6) = 0.$$
Factor out the common factors in each group:
$$2x(x - 2) + 3(x - 2) = 0.$$

Factor by grouping:
$$(x-2)(2x+3) = 0.$$

Setting each factor equal to zero gives:
$$x - 2 = 0 \quad \text{or} \quad 2x + 3 = 0.$$

Solving these, we obtain:
$$x = 2 \quad \text{or} \quad x = -\frac{3}{2}.$$

3. For the equation
$$4x^2 + 8x - 12 = 0,$$
notice that every term is divisible by 4. Factor out the common factor 4:
$$4(x^2 + 2x - 3) = 0.$$
Now, we factor the quadratic inside the parentheses. We need two numbers that multiply to -3 and add up to 2. The numbers 3 and -1 work because
$$3 \times (-1) = -3 \quad \text{and} \quad 3 + (-1) = 2.$$
Therefore,
$$x^2 + 2x - 3 = (x+3)(x-1).$$
Substituting back, we have:
$$4(x+3)(x-1) = 0.$$
By the zero product property, set each factor equal to zero:
$$x + 3 = 0 \quad \text{or} \quad x - 1 = 0.$$
This yields the solutions:
$$x = -3 \quad \text{or} \quad x = 1.$$

4. For the equation
$$3x^2 - 7x + 2 = 0,$$
compute the product $a \times c$ where $a = 3$ and $c = 2$:
$$3 \times 2 = 6.$$

We need two numbers that multiply to 6 and add to -7. The numbers -6 and -1 work since
$$(-6) \times (-1) = 6 \quad \text{and} \quad (-6) + (-1) = -7.$$
Rewrite the quadratic by splitting the middle term:
$$3x^2 - 6x - x + 2 = 0.$$
Group the terms:
$$(3x^2 - 6x) + (-x + 2) = 0.$$
Factor out the common factor from each group:
$$3x(x - 2) - 1(x - 2) = 0.$$
Factor out the common binomial factor:
$$(x - 2)(3x - 1) = 0.$$
Setting each factor equal to zero gives:
$$x - 2 = 0 \quad \text{or} \quad 3x - 1 = 0.$$
Thus, the solutions are:
$$x = 2 \quad \text{or} \quad x = \frac{1}{3}.$$

5. For the equation
$$6x^2 + 11x - 35 = 0,$$
first determine $a \times c$ where $a = 6$ and $c = -35$:
$$6 \times (-35) = -210.$$
Find two numbers that multiply to -210 and add to 11. The numbers 21 and -10 satisfy these conditions since
$$21 \times (-10) = -210 \quad \text{and} \quad 21 + (-10) = 11.$$
Split the middle term:
$$6x^2 + 21x - 10x - 35 = 0.$$

Group the terms:
$$(6x^2 + 21x) + (-10x - 35) = 0.$$

Factor out the common factors in each group:
$$3x(2x + 7) - 5(2x + 7) = 0.$$

Factor by grouping:
$$(2x + 7)(3x - 5) = 0.$$

Setting each factor equal to zero, we have:
$$2x + 7 = 0 \quad \text{or} \quad 3x - 5 = 0.$$

Solving these gives:
$$x = -\frac{7}{2} \quad \text{or} \quad x = \frac{5}{3}.$$

6. For the equation
$$x^2 - 6x + 9 = 0,$$
recognize that the quadratic is a perfect square trinomial. It can be written as
$$(x - 3)^2 = 0.$$
Taking the square root of both sides yields:
$$x - 3 = 0.$$
Therefore, the unique solution is:
$$x = 3.$$
This solution has a multiplicity of two.

Chapter 33

Completing the Square and the Quadratic Formula

Completing the Square

1 Theory and Method

A quadratic equation in standard form is expressed as

$$ax^2 + bx + c = 0,$$

where a, b, and c are constants with $a \neq 0$. The technique of completing the square transforms the quadratic expression into a perfect square trinomial, thereby enabling the equation to be solved in a systematic manner. When the leading coefficient is not unity, the first step is to divide the entire equation by a so that the quadratic term has a coefficient of 1:

$$x^2 + \frac{b}{a}x + \frac{c}{a} = 0.$$

The equation is then rearranged to isolate the constant term:

$$x^2 + \frac{b}{a}x = -\frac{c}{a}.$$

To create a perfect square trinomial on the left-hand side, it is necessary to add the square of one-half the coefficient of x to both

sides. That is, the quantity
$$\left(\frac{b}{2a}\right)^2$$
is added, yielding
$$x^2 + \frac{b}{a}x + \left(\frac{b}{2a}\right)^2 = -\frac{c}{a} + \left(\frac{b}{2a}\right)^2.$$
The left-hand side now factors neatly into
$$\left(x + \frac{b}{2a}\right)^2,$$
while the right-hand side can be written as
$$\frac{b^2 - 4ac}{4a^2}.$$
Taking the square root of both sides results in
$$x + \frac{b}{2a} = \pm\frac{\sqrt{b^2 - 4ac}}{2a}.$$
Isolating x by subtracting $\frac{b}{2a}$ then leads to the solutions of the quadratic equation.

2 Worked Example

Consider the quadratic equation
$$2x^2 + 8x + 6 = 0.$$
Dividing the entire equation by 2 transforms it into
$$x^2 + 4x + 3 = 0.$$
Rearrangement to isolate the constant term produces
$$x^2 + 4x = -3.$$
The next step involves calculating the square of half the coefficient of x. Since half of 4 is 2, its square is
$$2^2 = 4.$$

Adding 4 to both sides yields
$$x^2 + 4x + 4 = -3 + 4,$$
which simplifies to
$$(x+2)^2 = 1.$$
Extracting the square root of both sides gives
$$x + 2 = \pm 1.$$
Finally, subtracting 2 leads to the solutions:
$$x = -2 + 1 = -1 \quad \text{or} \quad x = -2 - 1 = -3.$$

The Quadratic Formula

1 Derivation from Completing the Square

Beginning with the standard quadratic equation
$$ax^2 + bx + c = 0,$$
and dividing through by a produces
$$x^2 + \frac{b}{a}x + \frac{c}{a} = 0.$$
Isolating the quadratic and linear terms gives
$$x^2 + \frac{b}{a}x = -\frac{c}{a}.$$
The addition of the square of half the coefficient of x, namely $\left(\frac{b}{2a}\right)^2$, to both sides results in
$$x^2 + \frac{b}{a}x + \left(\frac{b}{2a}\right)^2 = -\frac{c}{a} + \left(\frac{b}{2a}\right)^2.$$
The left-hand side factors as a perfect square:
$$\left(x + \frac{b}{2a}\right)^2 = \frac{b^2 - 4ac}{4a^2}.$$
Taking the square root of both sides leads to
$$x + \frac{b}{2a} = \pm\frac{\sqrt{b^2 - 4ac}}{2a}.$$
Subtracting $\frac{b}{2a}$ from both sides results in the quadratic formula:
$$x = \frac{-b \pm \sqrt{b^2 - 4ac}}{2a}.$$

2 Interpretation and Applications

The quadratic formula provides a direct method for solving any quadratic equation without reliance on factorization techniques that may not be readily apparent. The discriminant, defined as

$$\Delta = b^2 - 4ac,$$

plays a crucial role in determining the nature of the solutions. A positive discriminant indicates the presence of two distinct real roots, a discriminant of zero corresponds to a single repeated real root, and a negative discriminant implies that the equation has two complex conjugate roots. This formula encapsulates both the process of completing the square and the underlying structure of quadratic expressions, offering a comprehensive tool for analyzing and solving quadratic relationships in a variety of contexts.

Multiple Choice Questions

1. What is the first step when applying the method of completing the square to a quadratic equation where the coefficient of x^2 is not 1?

    (a) Multiply each term by a.
    (b) Divide the entire equation by a.
    (c) Subtract the constant term from both sides.
    (d) Add $\left(\frac{b}{2}\right)^2$ to both sides.

2. In the process of completing the square for a quadratic equation transformed into

$$x^2 + \frac{b}{a}x = -\frac{c}{a},$$

which term must be added to both sides to obtain a perfect square trinomial?

    (a) $\left(\frac{b}{2}\right)^2$
    (b) $\left(\frac{c}{2a}\right)^2$
    (c) $\left(\frac{b}{2a}\right)^2$
    (d) $\left(\frac{2a}{b}\right)^2$

3. After completing the square, the quadratic expression is rewritten as
$$\left(x + \frac{b}{2a}\right)^2 = -\frac{c}{a} + \left(\frac{b}{2a}\right)^2.$$
How can the right-hand side be simplified?

   (a) $\frac{b^2-4ac}{4a}$

   (b) $\frac{b^2+4ac}{4a^2}$

   (c) $\frac{b^2-4ac}{4a^2}$

   (d) $\frac{b^2+4ac}{4a}$

4. Which of the following expressions represents the quadratic formula derived from completing the square?

   (a) $x = \frac{-b \pm \sqrt{b^2-4ac}}{2}$

   (b) $x = \frac{-b \pm \sqrt{b^2-4ac}}{2a}$

   (c) $x = \frac{-b \pm \sqrt{b^2+4ac}}{2a}$

   (d) $x = \frac{-b \pm \sqrt{4ac-b^2}}{2a}$

5. The discriminant in the quadratic formula is given by
$$\Delta = b^2 - 4ac.$$
What information does the discriminant provide about the roots of the quadratic equation?

   (a) It indicates whether the quadratic expression can be factored over the integers.

   (b) It provides the sum of the roots.

   (c) It determines the number and type (real or complex) of the roots.

   (d) It represents the axis of symmetry of the quadratic graph.

6. In the worked example solving
$$2x^2 + 8x + 6 = 0,$$
the equation is first divided by 2 to yield
$$x^2 + 4x + 3 = 0,$$

368

and then rearranged to
$$x^2 + 4x = -3.$$
What number is added to both sides to complete the square?

(a) 1

(b) 2

(c) 3

(d) 4

7. Under which condition will the quadratic formula yield a pair of complex conjugate roots?

(a) When $b^2 - 4ac > 0$

(b) When $b^2 - 4ac = 0$

(c) When $b^2 - 4ac < 0$

(d) When $a = 0$

Answers:

1. **B: Divide the entire equation by a**
   Explanation: Dividing every term by a ensures that the coefficient of x^2 becomes 1, which is necessary for the procedure of completing the square.

2. **C: $\left(\frac{b}{2a}\right)^2$**
   Explanation: After dividing by a, the coefficient of x becomes $\frac{b}{a}$. To complete the square, you add the square of half this coefficient, which is $\left(\frac{b}{2a}\right)^2$.

3. **C: $\frac{b^2 - 4ac}{4a^2}$**
   Explanation: The term $\left(\frac{b}{2a}\right)^2$ simplifies to $\frac{b^2}{4a^2}$. When combined with $-\frac{c}{a}$ (rewritten as $-\frac{4ac}{4a^2}$), the right-hand side simplifies to $\frac{b^2 - 4ac}{4a^2}$.

4. **B: $x = \frac{-b \pm \sqrt{b^2 - 4ac}}{2a}$**
   Explanation: This is the standard form of the quadratic formula derived by completing the square on the general quadratic equation $ax^2 + bx + c = 0$.

5. **C: It determines the number and type (real or complex) of the roots**
   Explanation: The discriminant $b^2 - 4ac$ tells us whether the quadratic equation has two distinct real roots ($\Delta > 0$), one repeated real root ($\Delta = 0$), or two complex conjugate roots ($\Delta < 0$).

6. **D: 4**
   Explanation: For the expression $x^2 + 4x$, half of the coefficient of x is 2. Squaring 2 gives 4, which must be added to both sides to complete the square.

7. **C: When $b^2 - 4ac < 0$**
   Explanation: A negative discriminant implies that the term under the square root in the quadratic formula is negative, resulting in complex conjugate roots.

Practice Problems

1. Use the method of completing the square to solve the quadratic equation:
$$x^2 + 6x + 5 = 0.$$

2. Use the method of completing the square to solve the quadratic equation:
$$3x^2 + 6x - 9 = 0.$$

3. Derive the quadratic formula by completing the square on the standard form quadratic equation:

$$ax^2 + bx + c = 0.$$

4. Use the quadratic formula to solve the quadratic equation:

$$4x^2 - 12x + 9 = 0.$$

Then, interpret the value of the discriminant and explain the nature of the roots.

5. For the quadratic equation:

$$2x^2 - 3x + 1 = 0,$$

solve by both completing the square and using the quadratic formula. Compare the two methods and verify that they yield the same solutions.

6. Discuss the significance of the discriminant in the quadratic formula. Then, apply it to the quadratic equation:
$$x^2 + 4x + 4 = 0,$$
to determine the type and number of roots.

Answers

1. **Solution:**
   We begin with the quadratic equation:
   $$x^2 + 6x + 5 = 0.$$
   First, subtract 5 from both sides to isolate the quadratic and linear terms:
   $$x^2 + 6x = -5.$$
   Next, complete the square by adding the square of half the coefficient of x. Since half of 6 is 3, we add $3^2 = 9$ to both sides:
   $$x^2 + 6x + 9 = -5 + 9.$$
   This simplifies to:
   $$(x+3)^2 = 4.$$
   Taking the square root of both sides gives:
   $$x + 3 = \pm 2.$$
   Finally, subtract 3 from both sides to solve for x:
   $$x = -3 + 2 = -1 \quad \text{or} \quad x = -3 - 2 = -5.$$
   Therefore, the solutions are $x = -1$ and $x = -5$.

2. **Solution:**
   Consider the equation:
   $$3x^2 + 6x - 9 = 0.$$
   First, divide every term by 3 to simplify the equation:
   $$x^2 + 2x - 3 = 0.$$
   Rearrange the equation to separate the constant term:
   $$x^2 + 2x = 3.$$
   Next, complete the square by adding $\left(\frac{2}{2}\right)^2 = 1$ to both sides:
   $$x^2 + 2x + 1 = 3 + 1.$$
   The left-hand side factors as:
   $$(x+1)^2 = 4.$$
   Taking the square root of both sides yields:
   $$x + 1 = \pm 2.$$
   Solving for x, we find:
   $$x = -1 + 2 = 1 \quad \text{or} \quad x = -1 - 2 = -3.$$
   Thus, the solutions are $x = 1$ and $x = -3$.

3. **Solution:**
   Start with the standard quadratic equation:
   $$ax^2 + bx + c = 0.$$
   Divide every term by a (noting that $a \neq 0$) to obtain:
   $$x^2 + \frac{b}{a}x + \frac{c}{a} = 0.$$
   Isolate the quadratic and linear terms by subtracting $\frac{c}{a}$ from both sides:
   $$x^2 + \frac{b}{a}x = -\frac{c}{a}.$$

To complete the square, add the square of half the coefficient of x, which is $\left(\frac{b}{2a}\right)^2 = \frac{b^2}{4a^2}$, to both sides:

$$x^2 + \frac{b}{a}x + \frac{b^2}{4a^2} = -\frac{c}{a} + \frac{b^2}{4a^2}.$$

The left side factors to form a perfect square:

$$\left(x + \frac{b}{2a}\right)^2 = \frac{b^2 - 4ac}{4a^2}.$$

Taking the square root of both sides gives:

$$x + \frac{b}{2a} = \pm \frac{\sqrt{b^2 - 4ac}}{2a}.$$

Finally, solve for x by subtracting $\frac{b}{2a}$ from both sides:

$$x = \frac{-b \pm \sqrt{b^2 - 4ac}}{2a}.$$

This is the quadratic formula, derived by the method of completing the square.

4. **Solution:**
For the quadratic equation:

$$4x^2 - 12x + 9 = 0,$$

identify the coefficients as $a = 4$, $b = -12$, and $c = 9$. The quadratic formula is:

$$x = \frac{-b \pm \sqrt{b^2 - 4ac}}{2a}.$$

First, compute the discriminant:

$$b^2 - 4ac = (-12)^2 - 4(4)(9) = 144 - 144 = 0.$$

A discriminant of 0 indicates that there is exactly one real repeated root. Applying the quadratic formula:

$$x = \frac{-(-12) \pm \sqrt{0}}{2(4)} = \frac{12}{8} = \frac{3}{2}.$$

Thus, the equation has a single, repeated root $x = \frac{3}{2}$. The discriminant being zero also implies that the quadratic can be written as a perfect square, and geometrically, the parabola touches the x-axis at its vertex.

5. **Solution:**
   Consider the quadratic equation:
   $$2x^2 - 3x + 1 = 0.$$

   Using the Quadratic Formula:
   Here, $a = 2$, $b = -3$, and $c = 1$. The discriminant is:
   $$(-3)^2 - 4(2)(1) = 9 - 8 = 1.$$

   Applying the quadratic formula:
   $$x = \frac{-(-3) \pm \sqrt{1}}{2(2)} = \frac{3 \pm 1}{4}.$$

   This gives the solutions:
   $$x = \frac{3+1}{4} = 1 \quad \text{and} \quad x = \frac{3-1}{4} = \frac{1}{2}.$$

   Using Completing the Square:
   First, divide the entire equation by 2:
   $$x^2 - \frac{3}{2}x + \frac{1}{2} = 0.$$

   Rewrite the equation by moving the constant term:
   $$x^2 - \frac{3}{2}x = -\frac{1}{2}.$$

   To complete the square, add $\left(\frac{3}{4}\right)^2 = \frac{9}{16}$ to both sides:
   $$x^2 - \frac{3}{2}x + \frac{9}{16} = -\frac{1}{2} + \frac{9}{16}.$$

   Simplify the right side by writing $-\frac{1}{2}$ as $-\frac{8}{16}$:
   $$-\frac{8}{16} + \frac{9}{16} = \frac{1}{16}.$$

   The left-hand side is now a perfect square:
   $$\left(x - \frac{3}{4}\right)^2 = \frac{1}{16}.$$

   Taking the square root of both sides gives:
   $$x - \frac{3}{4} = \pm\frac{1}{4}.$$

Solving for x yields:
$$x = \frac{3}{4} \pm \frac{1}{4},$$
which results in:
$$x = 1 \quad \text{or} \quad x = \frac{1}{2}.$$
Both methods produce the same solutions, confirming the consistency of the approaches.

6. **Solution:**
The discriminant in the quadratic formula is given by:
$$\Delta = b^2 - 4ac.$$
Its value determines the nature of the roots of a quadratic equation:

- If $\Delta > 0$, there are two distinct real roots.
- If $\Delta = 0$, there is one real repeated root.
- If $\Delta < 0$, there are two complex conjugate roots.

Now, consider the quadratic equation:
$$x^2 + 4x + 4 = 0.$$
Here, $a = 1$, $b = 4$, and $c = 4$. Calculate the discriminant:
$$\Delta = 4^2 - 4(1)(4) = 16 - 16 = 0.$$
Since the discriminant is 0, the equation has one real repeated root. In fact, the quadratic factors neatly as:
$$(x + 2)^2 = 0,$$
leading to the solution:
$$x = -2.$$
This example illustrates that when the discriminant is 0, the graph of the quadratic function touches the x-axis at exactly one point—the vertex of the parabola.

Chapter 34

Mathematical Modeling Using Algebra

Foundations of Algebraic Modeling

Mathematical modeling transforms real-life situations into precise algebraic representations. Algebraic methods provide a systematic approach to express relationships found in physical, financial, and social contexts using equations and expressions. In this framework, natural phenomena and practical constraints are interpreted through symbols and operations, allowing abstract reasoning to be applied to concrete problems. Variables are introduced to stand for unknown quantities, thereby enabling the construction of equations that capture the inherent structure of a problem.

Defining Variables and Formulating Relationships

A central step in algebraic modeling involves the careful selection and definition of variables. The variables are chosen to correspond to quantities that are either unknown or subject to determination within the context of the problem. Once the relevant quantities are identified, relationships among these variables are expressed in the form of equations. For example, when modeling a scenario involving production costs and revenues, a variable may represent the number of units produced, while other expressions relate these units to overall profit. The process emphasizes clarity and consistency, ensuring that each variable is unambiguously tied to a specific quantity from the real-world situation being considered.

Developing Equations from Real-World Contexts

Translating practical scenarios into algebraic equations begins with the interpretation of descriptive statements into mathematical language. Phrases that imply operations such as addition, subtraction, multiplication, or division are converted into their symbolic counterparts. In many cases, the situation presents multiple constraints, leading to a system of equations that collectively represent the problem. For instance, in problems involving mixtures, if x and y denote the quantities of two solutions with differing concentrations, the total volume may be expressed as

$$x + y = T,$$

while a secondary equation accounting for the concentration relationship might take the form

$$c_1 x + c_2 y = C,$$

where T is the total volume, c and c are the concentrations of each solution, and C is the overall concentration of the mixture. Such equations encapsulate the relationships among the variables and reflect the logical structure inherent in the actual scenario.

Methods for Solving Algebraic Models

After establishing the equations that model a real-life situation, various algebraic techniques are employed to determine the values of the variables. Methods such as substitution, elimination, and factoring are used to solve systems of linear equations, while techniques including the quadratic formula and completing the square may be applied when the equations are nonlinear. Each approach not only yields numerical solutions but also enhances understanding of the behavior of the modeled system. The process of solving the equations reveals insights into how different factors interact and may indicate the presence of multiple solutions, some of which might need further interpretation when compared with the conditions imposed by the original problem.

Interpreting the Results of Algebraic Models

The final stage in the modeling process involves the interpretation of solutions within the context of the original problem. Once the values of the variables are determined, these numbers are mapped back to their corresponding real-world quantities. The interpretation step is crucial for verifying that the solutions are consistent with the practical constraints of the problem. For instance, if a variable represents a quantity that must be nonnegative, any negative solution would prompt a reexamination of the assumptions made during modeling. Furthermore, the numerical results provide insight into the relationships among the various components of the original scenario, reinforcing the idea that the algebraic model accurately reflects the underlying situation.

Detailed Examples in Algebraic Modeling

Numerous practical problems can be effectively addressed using algebraic modeling. In one situation, the revenue generated from the sale of a product can be modeled by linking the number of units sold with the sale price, resulting in an equation that describes total revenue. In another example, the motion of an object under uniform acceleration can be represented by equations relating displacement, velocity, and time. Additionally, modeling the growth or decline of a population within ecological studies often leads to equations that are either linear or quadratic. Each of these examples illustrates how algebra serves as a bridge between abstract mathematical principles and tangible real-world applications. The process involves identifying the critical quantities, forming the appropriate equations, applying systematic solution techniques, and finally interpreting the outcomes in a way that illuminates the original problem.

Multiple Choice Questions

1. Which of the following best describes the primary purpose of mathematical modeling using algebra?

    (a) To perform abstract calculations without real-world connections.

    (b) To transform real-life situations into precise algebraic representations.

    (c) To memorize formulas for solving standard equations.

    (d) To design computer algorithms solely based on numerical data.

2. In the process of algebraic modeling, why is it critical to carefully define variables?

    (a) Because they can be arbitrarily chosen without any significance.

    (b) Because they assign meaningful symbols to unknown or measurable quantities in the situation.

    (c) Because they help in completely avoiding equations.

    (d) Because they replace the need for developing relationships between quantities.

3. When translating a real-world problem into an equation for a mixture, which equation correctly represents the total volume of the two components?

    (a) $x - y = T$

    (b) $x \cdot y = Tx + y = T$

    (c) $x_{y=T}$

4. Which of the following methods is commonly used to solve systems of equations derived from algebraic models?

    (a) Substitution

    (b) Differentiation

    (c) Integration

    (d) Exponentiation

5. What is the main reason for interpreting the numerical solutions of an algebraic model back in the context of the original problem?

    (a) To verify that the correct formulas were used.
    (b) To ensure that the solutions are meaningful and satisfy any real-world constraints.
    (c) To simplify the algebraic expressions further.
    (d) To prepare the model for computer simulation.

6. Which scenario is most appropriately modeled using algebraic techniques?

    (a) Analyzing the trajectory of a projectile in physics.
    (b) Describing the plot of a novel.
    (c) Listing historical events chronologically.
    (d) Examining abstract art patterns.

7. If an algebraic model yields a negative solution for a variable that represents a quantity which must be nonnegative (such as population or volume), what should be the best course of action?

    (a) Accept the solution since all mathematically derived answers are valid.
    (b) Convert the negative value to positive by taking its absolute value.
    (c) Reexamine the model and assumptions to identify any errors or misinterpretations.
    (d) Discard the entire modeling process and start from scratch.

Answers:

1. **B: To transform real-life situations into precise algebraic representations**
   Explanation: Algebraic modeling is fundamentally about converting practical, real-world problems into mathematical expressions and equations that can be analyzed and solved.

2. **B: Because they assign meaningful symbols to unknown or measurable quantities in the situation**

Explanation: Defining variables appropriately is crucial because it establishes a clear connection between the symbols used in the equations and the real-world quantities they represent.

3. **C: x + y = T**
Explanation: In mixture problems, the total volume T is typically obtained by adding the individual volumes (x and y), making x + y = T the correct formulation.

4. **A: Substitution**
Explanation: Substitution is one of the common methods used to solve systems of equations in algebraic models, especially when one of the equations can be easily solved for one variable in terms of the others.

5. **B: To ensure that the solutions are meaningful and satisfy any real-world constraints**
Explanation: Interpreting the numerical results in the context of the original problem is essential to verify that the answers make sense (for example, ensuring a quantity that must be nonnegative indeed is nonnegative) and that the model is valid.

6. **A: Analyzing the trajectory of a projectile in physics**
Explanation: This scenario requires modeling relationships between variables such as displacement, velocity, and time, which is a classic application of algebraic techniques in a real-world context.

7. **C: Reexamine the model and assumptions to identify any errors or misinterpretations**
Explanation: If a variable that should only take on nonnegative values turns out negative, it suggests that there may be a flaw in the model or the assumptions made, prompting a review to ensure that the model accurately reflects the real-world scenario.

Practice Problems

1. Consider a scenario where a chemist wants to prepare 20 liters of a 25% salt solution by mixing a 10% salt solution with a 30% salt solution. Let x be the volume (in liters) of the 10%

solution and y be the volume of the 30% solution. Write the system of equations that models this situation.

2. A widget manufacturing company has a fixed cost of 1500 units and a variable cost of 5 units per widget. They sell each widget for 10 units.

> (a) Write an equation for the profit as
>
> a function of the number of widgets sold.
>
> (b) Determine the break-even point.

3. A textbook publisher plans to order textbooks and notebooks. The total number of items ordered is 200. Textbooks cost 8 units each while notebooks cost 3 units each, and the total cost is 1000 units. Let T represent the number of textbooks and N the number of notebooks. Formulate the system of equations representing this scenario.

4. A ball is thrown upward from an initial height of 5 feet with an initial velocity of 32 feet per second. Its height after t seconds is modeled by:

$$h(t) = -16t^2 + 32t + 5.$$

Determine the time at which the ball reaches its maximum height and compute that maximum height.

5. At a local theater, student tickets are sold for 8 units each and regular tickets for 12 units each. On a particular night, the theater sold 150 tickets and collected 1500 units in total. Let S be the number of student tickets sold and R be the number of regular tickets. Formulate the system of equations that models this scenario and solve for S and R.

6. Explain why it is important to interpret the solutions of an algebraic model in the context of the original real-world problem—especially when a solution yields a negative value for a quantity that must be nonnegative.

Answers

1. **Mixture Problem:**

   We let x be the volume (in liters) of the 10% solution and y be the volume (in liters) of the 30% solution. First, the total volume must add to 20 liters:
   $$x + y = 20.$$
   Next, the total amount of salt in the final mixture is 25% of 20 liters, i.e., 5 liters of salt. The salt from the 10% solution contributes 0.10x liters and the 30% solution contributes 0.30y liters, so:
   $$0.10x + 0.30y = 5.$$
   Explanation: The first equation enforces the volume constraint and the second equation ensures that the salt content matches the desired 25% concentration.

2. **Profit Equation and Break-even Analysis:**

   Let x represent the number of widgets sold. The revenue R(x) is:
   $$R(x) = 10x.$$
   The total cost C(x) is:
   $$C(x) = 1500 + 5x.$$
   Thus, the profit function P(x) is:
   $$P(x) = R(x) - C(x) = 10x - (1500 + 5x) = 5x - 1500.$$
   To find the break-even point, set P(x) equal to 0:
   $$5x - 1500 = 0 \quad \Rightarrow \quad 5x = 1500 \quad \Rightarrow \quad x = 300.$$
   Explanation: The break-even point occurs when revenue equals cost, meaning profit is zero. Here, selling 300 widgets covers both fixed and variable costs.

3. **Publisher Order Problem:**

   Let T be the number of textbooks and N be the number of notebooks. The total number of items is:
   $$T + N = 200.$$

The total cost is given by:
$$8T + 3N = 1000.$$

Explanation: The first equation ensures the correct item count, while the second accounts for the cost of each item type (textbooks at 8 units and notebooks at 3 units) to reach the total cost of 1000 units.

4. **Projectile Motion Problem:**

   The height function is:
   $$h(t) = -16t^2 + 32t + 5.$$

   The maximum height of a quadratic function occurs at the vertex, with t-coordinate given by:
   $$t = -\frac{b}{2a}$$

   where $a = -16$ and $b = 32$. Thus:
   $$t = -\frac{32}{2(-16)} = -\frac{32}{-32} = 1.$$

   Substitute $t = 1$ into h(t) to find the maximum height:
   $$h(1) = -16(1)^2 + 32(1) + 5 = -16 + 32 + 5 = 21.$$

   Explanation: Using the vertex formula, we find the ball reaches its maximum height at 1 second, and substituting back into the function gives the maximum height of 21 feet.

5. **Theater Ticket Problem:**

   Let S be the number of student tickets and R be the number of regular tickets. The total tickets equation is:
   $$S + R = 150.$$

   The revenue equation is:
   $$8S + 12R = 1500.$$

   Solve for S and R by first expressing S in terms of R:
   $$S = 150 - R.$$

Substitute into the revenue equation:
$$8(150 - R) + 12R = 1500.$$

Simplify:
$$1200 - 8R + 12R = 1500 \quad \Rightarrow \quad 1200 + 4R = 1500.$$

Solve for R:
$$4R = 1500 - 1200 = 300, \quad \text{so} \quad R = 75.$$

Then,
$$S = 150 - 75 = 75.$$

Explanation: The system is solved using substitution, revealing that 75 student tickets and 75 regular tickets were sold, satisfying both the total number of tickets and the total revenue collected.

6. **Interpreting Negative Solutions in Context:**

   In algebraic modeling, variables often represent quantities such as time, length, or counts, which cannot be negative in real-world contexts. If a solution to a modeled problem yields a negative value (for example, -20 tickets or -5 meters), this result is not meaningful in practice. It usually indicates that either the model needs adjustment or that the nonnegative constraint was not enforced during solution. By interpreting the solution within the original context, we ensure that only practical, feasible answers are accepted. **Explanation:** Always checking that solutions adhere to real-world constraints prevents the misuse of mathematically valid but practically invalid results. This step ensures that the algebraic model accurately reflects the situation being studied and maintains its relevance in real-life applications.

Chapter 35

Concept of Functions and Function Notation

Definition and Fundamental Characteristics of Functions

A function is a mathematical relation that establishes a correspondence between elements of one set, known as the domain, and elements of another set, often referred to as the codomain, in such a manner that each element in the domain is associated with exactly one element in the codomain. This relation is commonly expressed by a symbol such as f and written in function notation as $f : A \to B$, where A represents the domain and B represents the codomain. The notation $f(x)$ signifies the output corresponding to an input x in A, making the concept of input-output association both clear and concise. Such precision in definitions is essential in ensuring that the relationship between varying quantities is unambiguous and mathematically sound.

Domain of a Function

The domain of a function is defined as the set of all possible input values for which the function yields a meaningful result. Determining the domain involves careful consideration of the operations present in the functional expression. For instance, when a function contains a square root, the expression inside the radical must be nonnegative; an inequality such as $x - 2 \geq 0$ may arise in a function defined by $f(x) = \sqrt{x-2}$, thereby restricting the domain to the interval $[2, \infty)$. In the case of functions expressed as rational expressions, the denominator must be nonzero, meaning that any value of x which would lead to division by zero is excluded from the domain. This systematic analysis ensures that the function is defined only for those inputs that produce valid and real outputs.

Range of a Function

The range of a function comprises all output values that result from evaluating the function over its domain. In other words, while the domain indicates the permissible inputs, the range reflects the collection of outputs that the function is capable of producing. For example, consider the function defined by $f(x) = x^2$; although the domain may include all real numbers, the nature of the squaring operation ensures that every output is nonnegative. Accordingly, the range of this function is described by the set

$$\{y \in \mathbb{R} \mid y \geq 0\}.$$

A thorough understanding of the range involves analyzing how the function transforms the domain and noting any inherent restrictions that limit the output values. This concept is crucial in understanding the overall behavior of the function and the scope of its application.

Proper Function Notation

The conventional notation used to denote functions is both precise and highly informative. Typically, a function is identified by a letter, such as f, followed by its independent variable in parentheses; for instance, $f(x)$ explicitly shows that the function f depends on the variable x. In a more formal setting, a function is defined by an expression such as

$$f : A \to B, \quad \text{with} \quad f(x) = \text{expression in } x,$$

where A denotes the domain and B denotes the codomain. This notation not only clarifies the relationship between inputs and outputs but also provides a framework for further operations, such as function composition or the determination of inverse functions.

The unambiguous use of symbols and notation ensures that the mathematical description of relationships remains consistent and is readily interpretable, thereby constituting a fundamental component in the study and application of algebraic concepts.

Multiple Choice Questions

1. Which of the following best describes a function?

    (a) A relation that assigns each element of the codomain to exactly one element of the domain.

    (b) A relation that assigns each element of the domain to exactly one element of the codomain.

    (c) A relation that allows an element in the domain to be mapped to multiple elements in the codomain.

    (d) A relation where every element of the domain is paired with every element of the codomain.

2. In the notation $f : A \to B$, what does the set A represent?

    (a) The codomain.

    (b) The range.

    (c) The domain.

    (d) The set of output values.

3. Consider the function $f(x) = \sqrt{x-2}$. Which of the following correctly describes its domain?

    (a) $(-\infty, 2]$

    (b) $[2, \infty)$

    (c) All real numbers.

    (d) $(2, \infty)$

4. Which of the following statements best describes the range of a function?

    (a) The set of all input values for which the function is defined.

    (b) The rule that assigns an output to each input.

    (c) The set of all output values resulting from the function.

(d) The expression obtained by substituting x into the function.

5. Which notation properly conveys the definition of a function?

    (a) $f = \{x, f(x)\}$
    (b) $f : A \to B, \quad f(x) = $ expression in x
    (c) $f(x) = f$ with $x \in A$
    (d) $f : B \to A, \quad f(x) = $ expression in x

6. If a function is defined by $f(x) = x^2$ with $f : \mathbb{R} \to \mathbb{R}$, what is its range?

    (a) All real numbers.
    (b) $[0, \infty)$
    (c) $(-\infty, \infty)$
    (d) $(-\infty, 0]$

7. Which statement is true regarding the mapping of a function?

    (a) Every element in the codomain must be mapped to by at least one element in the domain.
    (b) An element in the domain can be associated with more than one element in the codomain.
    (c) Each element in the domain is mapped to exactly one element in the codomain.
    (d) The domain and codomain are always identical.

Answers:

1. **B: A relation that assigns each element of the domain to exactly one element of the codomain**
   A function, by definition, ensures that every input from the domain is associated with exactly one output in the codomain. Option B captures this one-to-one mapping characteristic.

2. **C: The domain**
   In the function notation $f : A \to B$, the set A represents the domain—i.e., the set of all permissible input values for the function.

3. **B:** $[2, \infty)$
   Since the function $f(x) = \sqrt{x-2}$ involves a square root, the expression inside the radical must be nonnegative; setting $x - 2 \geq 0$ yields $x \geq 2$, which corresponds to the interval $[2, \infty)$.

4. **C: The set of all output values resulting from the function**
   The range of a function consists of all outputs generated when the function is applied to its domain. Option C correctly defines the range as the set of all such outputs.

5. **B:** $f : A \to B, \quad f(x) = $ **expression in** x
   This notation explicitly identifies the function's domain A, codomain B, and the rule $f(x)$ that assigns each input x to an output. It is the conventional and precise way to define a function.

6. **B:** $[0, \infty)$
   For the function $f(x) = x^2$ defined on all real numbers, squaring produces nonnegative outputs only. Even though every real number is a valid input, the outputs are always ≥ 0, so the range is $[0, \infty)$.

7. **C: Each element in the domain is mapped to exactly one element in the codomain**
   This statement encapsulates the core property of functions: every input from the domain has one and only one corresponding output in the codomain. Options A, B, and D do not accurately reflect this definition.

Practice Problems

1. Determine whether the relation

   $$R = \{(1,3),\ (2,5),\ (1,7),\ (3,9)\}$$

   is a function. Explain your reasoning.

2. Determine the domain of the function:
$$f(x) = \sqrt{3-x}$$

3. Determine the range of the function:
$$f(x) = x^2,$$
where the domain of f is all real numbers.

4. Express in proper function notation the following description: "Associate each real number with its cube minus one."

5. Determine the domain of the function:
$$f(x) = \frac{1}{x-5}$$

6. Explain the difference between the codomain and the range of a function. Provide an example to illustrate your explanation.

Answers

1. **Solution:** A relation is a function if every input is associated with exactly one output. In the relation
$$R = \{(1,3),\ (2,5),\ (1,7),\ (3,9)\},$$
notice that the input 1 appears twice with two different outputs (3 and 7). Since one input (1) is related to more than one output, the relation does not satisfy the definition of a function.

2. **Solution:** For the function
$$f(x) = \sqrt{3-x},$$

the expression under the square root (the radicand) must be nonnegative. Therefore, we require:
$$3 - x \geq 0.$$
Solving this inequality:
$$x \leq 3.$$
Hence, the domain of the function is all real numbers less than or equal to 3, which is written as:
$$(-\infty, 3].$$

3. **Solution:** Consider the function
$$f(x) = x^2,$$
defined for all real numbers. Since squaring any real number always yields a nonnegative result, the output of the function is never less than 0. Thus, the range of the function is all nonnegative real numbers:
$$[0, \infty).$$

4. **Solution:** The description "Associate each real number with its cube minus one" can be written in proper function notation by first specifying the domain and codomain, and then defining the rule of assignment. One acceptable notation is:
$$f : \mathbb{R} \to \mathbb{R}, \quad f(x) = x^3 - 1.$$
This notation indicates that the function f takes any real number x and maps it to $x^3 - 1$.

5. **Solution:** For the function
$$f(x) = \frac{1}{x - 5},$$
the denominator must not be zero, since division by zero is undefined. Therefore, we require:
$$x - 5 \neq 0,$$
which simplifies to:
$$x \neq 5.$$
Thus, the domain of the function is all real numbers except 5. In interval notation, this is:
$$(-\infty, 5) \cup (5, \infty).$$

6. **Solution:** The **codomain** of a function is the set of all values that could possibly come out as outputs according to the function's definition, while the **range** (or image) is the set of values that the function actually produces when applied to every element in its domain.

For example, consider the function defined by:
$$f : \mathbb{R} \to \mathbb{R}, \quad f(x) = x^2.$$

Here, the codomain is $\mathbb{R}$ (the set of all real numbers) because that is what is specified in the function's definition. However, since squaring any real number results in a nonnegative number, the function only outputs values in the interval:
$$[0, \infty),$$
which is the range. This example clearly shows that although the codomain is all of $\mathbb{R}$, the range of the function is limited to the nonnegative real numbers.

Chapter 36

Graphing and Analyzing Functions

Fundamentals of Graphing Functions

Every function can be represented as a collection of ordered pairs in the Cartesian coordinate system, where each point is of the form $(x, f(x))$. The horizontal axis represents the independent variable x, while the vertical axis represents the dependent variable $f(x)$. Graphing a function requires plotting these points accurately and connecting them in a manner that reflects the continuity or discreteness inherent in the function. Detailed analysis of precise points, symmetry, and curvature plays a significant role in understanding the behavior of the graph. The proper choice of scale and the identification of key points are essential for capturing the distinctive features of the function.

Transformations of Functions

Transformations modify the graph of a function through systematic alterations of its equation. A transformation may include translations, reflections, dilations, or compressions, each of which affects the appearance of the graph without altering the underlying relationship between the variables.

A horizontal translation is represented by an equation of the form
$$g(x) = f(x - h),$$

where shifting the graph to the right occurs when $h > 0$ and to the left when $h < 0$. Similarly, a vertical translation is captured by

$$g(x) = f(x) + k,$$

where a positive value of k shifts the graph upward, and a negative value shifts it downward.

Reflections occur when the sign of either the input or the output is reversed. For instance, the function

$$g(x) = f(-x)$$

produces a reflection of the graph of f across the y-axis, whereas

$$g(x) = -f(x)$$

results in a reflection across the x-axis. Dilation involves stretching or compressing the graph. A vertical stretch or compression is expressed as

$$g(x) = a\,f(x),$$

where the magnitude of a determines the degree of stretching (if $|a| > 1$) or compression (if $0 < |a| < 1$). A horizontal stretch or compression may be represented by

$$g(x) = f(bx),$$

with similar conditions on the value of $|b|$ affecting the transformation.

Intercepts and Key Graphical Features

Intercepts serve as anchor points that facilitate the accurate plotting of a function. The x-intercepts are found by solving the equation

$$f(x) = 0,$$

which identifies the points where the graph crosses or touches the x-axis. In contrast, the y-intercept is determined by evaluating the function at $x = 0$, provided that zero lies within the domain of the function. The coordinates of the y-intercept are given by

$$(0, f(0)).$$

Additional key features of the graph include points of inflection, relative maximums, and relative minimums. These points can be located through analytical methods and are important in outlining the general shape and characteristics of the function. A systematic algebraic procedure, followed by careful plotting, reveals how these features define the overall geometry of the graph.

Analyzing the Overall Shape of Functions

The overall shape of a function's graph is influenced by a combination of its algebraic form and the transformations previously described. For many functions, symmetry plays an important role. A function that satisfies the condition

$$f(-x) = f(x)$$

exhibits symmetry with respect to the y-axis, creating a mirror image on either side of the axis. Alternatively, if the function satisfies

$$f(-x) = -f(x),$$

the graph is symmetric with respect to the origin, indicating rotational symmetry.

The behavior at extreme values of x, known as end behavior, also contributes to the overall shape. In polynomial functions, the degree of the polynomial and the sign of its leading coefficient determine how the graph behaves as x approaches positive or negative infinity. In functions involving asymptotic behavior, the presence of horizontal or vertical asymptotes governs the approach of the curve to specific lines, without ever intersecting them.

Curvature, including regions of concavity and convexity, is another aspect of the graph's shape that provides insight into the rate of change of the function. Detailed examination of these properties involves the study of derivatives, which describe the instantaneous rate of change, thereby indicating intervals where the function is increasing or decreasing. Graphical analysis also includes the identification of turning points that indicate transitions from rising to falling values or vice versa.

Collectively, the techniques of graphing and the analytical investigation of transformations, intercepts, and overall shape form a comprehensive framework for understanding the behavior of functions.

Multiple Choice Questions

1. Which of the following best represents a point on the graph of a function f?

    (a) (f(x), x)
    (b) (x, f(x))
    (c) (f(x), f(x))
    (d) (x, x)

2. The transformation defined by
$$g(x) = f(x - h)$$
produces which effect on the graph of f when $h > 0$?

    (a) The graph shifts upward by h units.
    (b) The graph shifts to the right by h units.
    (c) The graph shifts to the left by h units.
    (d) The graph shifts downward by h units.

3. For the transformation given by
$$g(x) = -f(x),$$
what change occurs to the graph of f?

    (a) It is reflected over the y-axis.
    (b) It undergoes a horizontal stretch.
    (c) It is reflected over the x-axis.
    (d) It is shifted vertically.

4. How is the y-intercept of a function f determined when graphing it?

    (a) By solving the equation $f(x) = 0$ for x.
    (b) By evaluating the function at $x = 0$.
    (c) By finding the maximum value of the function.
    (d) By setting $f(x) = x$ and solving.

5. A function that satisfies the condition
$$f(-x) = f(x)$$
exhibits which type of symmetry?

    (a) Symmetry with respect to the x-axis.
    (b) Symmetry with respect to the origin.
    (c) Symmetry with respect to the y-axis.
    (d) No symmetry.

6. Consider the transformation
$$g(x) = f(bx)$$
where $|b| > 1$. What is the effect on the graph of f?

    (a) It undergoes a horizontal stretch.
    (b) It undergoes a horizontal compression.
    (c) It undergoes a vertical stretch.
    (d) It undergoes a vertical compression.

7. Which analytical tool is most useful for determining the intervals on which a function is increasing or decreasing?

    (a) Identifying the intercepts.
    (b) Analyzing the points of symmetry.
    (c) Computing the derivative of the function.
    (d) Calculating the y-intercept.

Answers:

1. **B: (x, f(x))**
   Every point on the graph of a function is represented as an ordered pair, where the first component is the input x and the second component is its corresponding output $f(x)$.

2. **B: The graph shifts to the right by h units.**
   The function $g(x) = f(x - h)$ causes a horizontal translation. When $h > 0$, the graph of f moves to the right by h units.

3. **C: It is reflected over the x-axis.**
   Multiplying the function by -1 as in $g(x) = -f(x)$ inverts all the y-values, reflecting the graph across the x-axis.

4. **B: By evaluating the function at $x = 0$.**
   The y-intercept is found by substituting $x = 0$ into the function, yielding the point $(0, f(0))$, provided that 0 is in the function's domain.

5. **C: Symmetry with respect to the y-axis.**
   If $f(-x) = f(x)$, the graph mirrors itself on the left and right of the y-axis, indicating y-axis symmetry.

6. **B: It undergoes a horizontal compression.**
   In the transformation $g(x) = f(bx)$, when $|b| > 1$, the graph of f is compressed horizontally by a factor of $\frac{1}{|b|}$.

7. **C: Computing the derivative of the function.**
   The derivative provides the instantaneous rate of change, which is used to determine the intervals over which the function is increasing or decreasing.

Practice Problems

1. Graph the quadratic function
$$f(x) = x^2 - 4.$$
Identify its x-intercepts, y-intercept, vertex, and line of symmetry.

2. Consider the function
$$f(x) = \sqrt{x}.$$
Plot the function and determine its domain, intercept(s), and end behavior. Describe any key features of its graph.

3. Let
$$f(x) = x^3.$$
First, reflect the graph of $f(x)$ across the x-axis, and then shift it downward by 2 units. Write the equation of the transformed function and explain how the graph is altered.

4. For the linear function
$$g(x) = 3x - 6,$$
find both the x-intercept and the y-intercept. Explain how these intercepts assist in graphing the line.

5. Consider the quadratic function
$$f(x) = (x - 2)^2.$$
Determine the vertex and the axis of symmetry from its equation. Discuss why these features are important for graphing the parabola.

6. The exponential function
$$f(x) = 2^x$$
undergoes a horizontal compression by a factor of 2 and a vertical stretch by a factor of 3. Write the equation of the transformed function and discuss its asymptotic behavior following these transformations.

Answers

1. **Solution:** For the function
$$f(x) = x^2 - 4,$$
the x-intercepts are found by setting
$$x^2 - 4 = 0.$$
Solving, we have
$$x^2 = 4 \implies x = \pm 2.$$
Therefore, the x-intercepts are $(-2, 0)$ and $(2, 0)$. The y-intercept is found by evaluating $f(0)$:
$$f(0) = 0^2 - 4 = -4,$$
so the y-intercept is $(0, -4)$. To find the vertex, note that the equation is already in the form
$$f(x) = x^2 - 4,$$
which can be written as
$$(x - 0)^2 - 4.$$

Hence, the vertex is at $(0, -4)$. The parabola is symmetric about the vertical line $x = 0$ (the y-axis). This analysis shows that the graph is an upward-opening parabola with its lowest point at $(0, -4)$ and intercepts at $(-2, 0)$ and $(2, 0)$.

2. **Solution:** For
$$f(x) = \sqrt{x},$$
the domain is determined by the requirement that the radicand be nonnegative:
$$x \geq 0.$$
The y-intercept is found by evaluating $f(0)$:
$$f(0) = \sqrt{0} = 0,$$
so the graph passes through $(0, 0)$. Since the function is only defined for $x \geq 0$, the x-intercept is also $(0, 0)$. As x increases, $\sqrt{x}$ increases gradually; its rate of increase slows for larger x. There is no upper bound, so as $x \to \infty$, $f(x) \to \infty$. Key features of the graph include its starting point at the origin, a smoothly increasing curve for $x > 0$, and the fact that it is not defined for negative values of x.

3. **Solution:** Starting with
$$f(x) = x^3,$$
a reflection across the x-axis is achieved by multiplying the entire function by -1, giving
$$g(x) = -x^3.$$
Next, a downward shift by 2 units is performed by subtracting 2 from the function:
$$g(x) = -x^3 - 2.$$
In the original function x^3, the graph has rotational symmetry about the origin and increases without bound as x increases. The reflection across the x-axis inverts the graph so that the portions which were above the x-axis now lie below, and vice versa. The subsequent shift downward moves every point on the graph 2 units lower along the y-axis. This transformation alters the location of key points such as the inflection point (which shifts from $(0, 0)$ to $(0, -2)$) while preserving the overall cubic shape.

4. **Solution:** For
$$g(x) = 3x - 6,$$
the y-intercept is determined by setting $x = 0$:
$$g(0) = 3(0) - 6 = -6,$$
which results in the point $(0, -6)$. The x-intercept is found by solving
$$3x - 6 = 0.$$
Adding 6 to both sides gives:
$$3x = 6,$$
hence
$$x = 2.$$
Therefore, the x-intercept is at $(2, 0)$. These intercepts are crucial because they provide two fixed points through which the straight line passes, making it straightforward to graph the function by drawing a line through $(0, -6)$ and $(2, 0)$.

5. **Solution:** The quadratic function
$$f(x) = (x - 2)^2$$
is given in vertex form. The general vertex form is
$$f(x) = (x - h)^2 + k,$$
where (h, k) is the vertex. Here, $h = 2$ and $k = 0$, so the vertex is $(2, 0)$. The axis of symmetry is the vertical line passing through the vertex, which is
$$x = 2.$$
Identifying the vertex and axis of symmetry is important as they indicate the minimum (or maximum) point of the parabola and allow the graph to be reflected evenly on both sides of the axis $x = 2$, leading to an accurate sketch of the parabola.

6. **Solution:** For the exponential function
$$f(x) = 2^x,$$

a horizontal compression by a factor of 2 means that every x-value is effectively doubled inside the function argument. This is achieved by replacing x with $2x$, yielding

$$f_1(x) = 2^{2x}.$$

A vertical stretch by a factor of 3 is then applied by multiplying the function by 3:

$$g(x) = 3 \cdot 2^{2x}.$$

Thus, the equation of the transformed function is

$$g(x) = 3 \cdot 2^{2x}.$$

Despite these transformations, the horizontal asymptote remains $y = 0$ because multiplying by a constant and compressing horizontally does not alter the end behavior as $x \to -\infty$. However, as $x \to \infty$ the function grows much more rapidly due to the horizontal compression, and the vertical stretch amplifies all output values by a factor of 3. This transformation results in an exponential growth curve that rises sharply for positive x while still approaching zero for very negative x.

Chapter 37

Linear Functions and Their Applications

Definition and Fundamental Characteristics

A linear function is represented by an equation of the form

$$f(x) = mx + b,$$

where m denotes the slope and b represents the y-intercept. The slope m measures the constant rate of change of the function, indicating how the dependent variable $f(x)$ changes for each unit change in the independent variable x. The y-intercept b is the value of the function when $x = 0$ and serves as the point where the graph crosses the y-axis. This inherent linearity implies that the first derivative is constant, and the graph of the function always forms a straight line in the Cartesian coordinate system.

Graphical Representation and Analysis

Graphing a linear function involves plotting points that satisfy the relation defined by the equation $f(x) = mx + b$. The graph of a linear function is a straight line, uniquely determined by its slope and y-intercept. In the Cartesian plane, the y-intercept appears at the coordinate $(0, b)$. When the slope m is nonzero, the x-intercept

can be determined by solving
$$mx + b = 0,$$
which yields
$$x = -\frac{b}{m}.$$
The constant slope ensures that the distance between any two points on the line maintains a fixed ratio of vertical change to horizontal change. Analysis of the graph emphasizes the importance of these intercepts and the uniform gradient, revealing the predictable behavior of linear functions over their entire domain.

Algebraic Forms of Linear Equations

Linear equations may be expressed in several equivalent forms, each providing certain advantages for analysis. The slope-intercept form,
$$y = mx + b,$$
directly displays the slope and y-intercept, enabling immediate identification of these key characteristics. In situations where a specific point on the line is known, the point-slope form,
$$y - y_1 = m(x - x_1),$$
proves invaluable, as it utilizes the known coordinates (x_1, y_1) along with the slope m to construct the equation. An alternative presentation is the standard form,
$$Ax + By = C,$$
which is particularly useful for solving systems of linear equations or for applications requiring a balanced representation of both variables. The multiple representations preserve the linearity of the relationship while affording flexibility in problem-solving and analytical contexts.

Practical Applications in Problem Solving

Linear functions offer a robust framework for modeling and solving real-world problems where relationships between variables exhibit a constant rate of change. In economics, for instance, cost functions are frequently modeled by linear equations that combine fixed costs with variable costs proportional to production levels. Similarly, in physics, the relationship between distance and time for an object moving at constant speed is governed by a linear function, with the slope representing speed and the y-intercept indicating initial displacement. Formulating these models involves identifying the independent and dependent variables, determining the rate of change, and setting up an equation that accurately reflects the situation. The properties of linear functions then facilitate the systematic solution of the problem by allowing isolation of the variable of interest and clear interpretation of the relationships involved.

Interpreting Slope and Y-Intercept in Real-World Contexts

The slope and y-intercept of a linear function carry significant interpretive meaning in practical applications. The value of the slope m symbolizes the rate at which the response variable changes per unit variation in the predictor variable. A positive slope indicates a direct relationship, where the response increases as the predictor increases, while a negative slope suggests an inverse relationship. The y-intercept b often corresponds to an initial or baseline condition when the independent variable is zero. In applied contexts such as budgeting or forecasting, the y-intercept may represent a fixed starting value such as an initial cost or base amount, whereas the slope reveals the incremental changes associated with additional units. Analyzing these components within the framework of a linear model yields a comprehensive understanding of the underlying phenomena and supports the effective application of mathematical reasoning to practical problem-solving.

Multiple Choice Questions

1. Which of the following equations correctly represents the slope-intercept form of a linear function?

    (a) f(x) = mx + b

    (b) $f(x) = ax^2 + bx + c f(x) = (x - h)^2 + k$

(c) $f(x) = m\overline{x+b}$

2. In the linear function f(x) = mx + b, what does the parameter m represent?

    (a) The y-intercept of the function
    (b) The slope or rate of change of the function
    (c) The x-intercept of the function
    (d) The vertical shift of the graph

3. To find the x-intercept of a linear function given in slope-intercept form, which step is required?

    (a) Set x = 0 and solve for y.
    (b) Set f(x) = 0 and solve for x.
    (c) Use the value of b directly as the x-intercept.
    (d) Find the derivative and set it equal to zero.

4. Which form of a linear equation is most convenient when you know a specific point on the line and its slope?

    (a) Slope-intercept form
    (b) Point-slope form
    (c) Standard form
    (d) Intercept form

5. In a cost function modeled by the linear equation C(x) = mx + b, where x represents the number of items produced, what does the constant term b typically represent?

    (a) The variable cost per unit
    (b) The fixed or initial cost regardless of production level
    (c) The marginal cost of each additional unit
    (d) The total revenue when no items are produced

6. What does a negative value for the slope m indicate about the relationship between x and f(x) in a linear function?

    (a) f(x) remains constant as x increases.
    (b) f(x) increases as x increases.
    (c) f(x) decreases as x increases.
    (d) f(x) exhibits a parabolic behavior.

7. Why is the graph of any linear function always a straight line on the Cartesian plane?

    (a) Because it has a constant rate of change and a constant first derivative.

    (b) Because it always passes through the origin.

    (c) Because the y-intercept is always nonzero.

    (d) Because the x-intercept is always equal to zero.

Answers:

1. **A: f(x) = mx + b**
   This is the slope-intercept form where m represents the slope and b represents the y-intercept. The other options represent different types of functions or incorrect forms.

2. **B: The slope or rate of change of the function**
   In $f(x) = mx + b$, m is the slope, indicating how much f(x) changes for a unit change in x. It defines the constant rate of change of the linear function.

3. **B: Set f(x) = 0 and solve for x**
   The x-intercept occurs where $f(x) = 0$. Setting $mx + b = 0$ and solving for x (i.e., x = -b/m) yields the x-intercept.

4. **B: Point-slope form**
   The point-slope form, given by $y - y_1 = m(x - x_1)$, is especially useful when you know one point on the line and the slope, allowing easy formulation of the line's equation.

5. **B: The fixed or initial cost regardless of production level**
   In cost functions, the constant term b typically represents the fixed cost (or initial cost) that is incurred even when no units are produced, while m represents the variable cost per unit.

6. **C: f(x) decreases as x increases**
   A negative slope indicates that as the independent variable x increases, the dependent variable f(x) decreases, leading to a downward-trending line.

7. **A: Because it has a constant rate of change and a constant first derivative**

A linear function is a first-degree polynomial; it has a constant slope (first derivative) which ensures that its graph is a straight line across the Cartesian plane.

Practice Problems

1. Consider the linear function
$$f(x) = 3x - 9.$$
Identify the slope and y-intercept, and then determine the x-intercept of the line.

2. Convert the standard form equation
$$2x + 5y = 10$$
into slope-intercept form. State the slope and y-intercept of the resulting equation.

3. Write the equation of a line in point-slope form that passes through the point $(4, -2)$ with a slope of 3.

4. Find the equation of the line in slope-intercept form that passes through the points
$$(1, 2)$$
and
$$(5, 10).$$

5. A taxi company charges a fixed fee of
$$4$$
units plus
$$1.50$$
units per mile traveled. Construct a linear function
$$F(m)$$
that represents the total fare for a ride covering
$$m$$
miles, and calculate the fare for a 15-mile trip.

6. The linear function
$$C(x) = 0.75x + 20$$
represents the total cost in dollars of buying x items, where the fixed cost includes a shipping fee. Explain what the slope and y-intercept represent in this context.

Answers

1. For the function
$$f(x) = 3x - 9,$$
the slope m is the coefficient of x, which is 3. The y-intercept b is -9, meaning the line crosses the y-axis at $(0, -9)$. To find the x-intercept, set $f(x) = 0$:
$$3x - 9 = 0 \quad \Rightarrow \quad 3x = 9 \quad \Rightarrow \quad x = 3.$$
Thus, the x-intercept is $(3, 0)$.

2. Starting with the standard form equation
$$2x + 5y = 10,$$
subtract $2x$ from both sides to isolate the term with y:
$$5y = -2x + 10.$$
Divide both sides by 5:
$$y = -\frac{2}{5}x + 2.$$
Therefore, the slope is $-\frac{2}{5}$ and the y-intercept is 2, meaning the line crosses the y-axis at $(0, 2)$.

3. The point-slope form of a line is given by:
$$y - y_1 = m(x - x_1).$$
For the point $(4, -2)$ and slope $m = 3$, substitute into the formula:
$$y - (-2) = 3(x - 4) \quad \Rightarrow \quad y + 2 = 3(x - 4).$$
This is the equation of the line in point-slope form.

4. To find the equation of the line that passes through $(1, 2)$ and $(5, 10)$, first compute the slope m:
$$m = \frac{10 - 2}{5 - 1} = \frac{8}{4} = 2.$$
Next, use the slope-intercept form $y = mx + b$. Substitute the point $(1, 2)$ into the equation to solve for b:
$$2 = 2(1) + b \quad \Rightarrow \quad b = 2 - 2 = 0.$$
Thus, the equation of the line is:
$$y = 2x.$$

5. The taxi fare is modeled by the linear function where the fixed fee is the y-intercept and the per-mile charge is the slope. Therefore, the function is:
$$F(m) = 1.50m + 4.$$
For a 15-mile trip, substitute $m = 15$:
$$F(15) = 1.50(15) + 4 = 22.5 + 4 = 26.5.$$
Hence, the fare for a 15-mile trip is 26.5 units.

6. In the function
$$C(x) = 0.75x + 20,$$
the slope 0.75 represents the cost per item; that is, for every additional item purchased, the total cost increases by 0.75 dollars. The y-intercept 20 represents the fixed cost (such as a shipping fee) that is incurred even if no items are bought. Therefore, when $x = 0$, the cost is 20 dollars solely from the fixed fee.

Chapter 38

Operations and Composition of Functions

Arithmetic Operations on Functions

Arithmetic operations provide a structured method for combining functions to generate new functions. Given two functions, f and g, various operations can be performed on them, each defined by the operation applied to the outputs of the functions for the same input.

1 Addition and Subtraction of Functions

For functions f(x) and g(x), the sum and difference are defined by carrying out the arithmetic operations on their values at any input x. In particular, the sum of the functions is given by

$$(f+g)(x) = f(x) + g(x),$$

and the difference is defined as

$$(f-g)(x) = f(x) - g(x).$$

It is important to note that the domain of the resulting function consists of all x-values that belong to both the domain of f and the domain of g. For example, if

$$f(x) = 2x + 3 \quad \text{and} \quad g(x) = x^2,$$

then their sum becomes

$$(f + g)(x) = 2x + 3 + x^2,$$

which is valid for all x that are allowable in both f and g.

2 Multiplication and Division of Functions

Multiplication of functions involves multiplying their outputs for each x. Formally, the product of f and g is expressed as

$$(f \cdot g)(x) = f(x) \cdot g(x).$$

Division of functions is handled similarly over the domain where the divisor is nonzero. The quotient is defined as

$$\left(\frac{f}{g}\right)(x) = \frac{f(x)}{g(x)},$$

with the stipulation that $g(x) \neq 0$ in the domain. As an example, if

$$f(x) = x - 1 \quad \text{and} \quad g(x) = x + 2,$$

then the product is

$$(f \cdot g)(x) = (x - 1)(x + 2),$$

and the quotient is

$$\left(\frac{f}{g}\right)(x) = \frac{x - 1}{x + 2},$$

with the condition that $x \neq -2$.

Composition of Functions

Composition of functions creates a new function by applying one function to the result of another. Given functions f and g, the composition f∘g is defined by first applying g to an input x and then applying f to the outcome of g(x). This operation is written as

$$(f \circ g)(x) = f\bigl(g(x)\bigr).$$

Similarly, the composition g∘f is defined by
$$(g \circ f)(x) = g(f(x)).$$

It is noteworthy that, in general, function composition is not commutative; that is, $(f \circ g)(x)$ usually differs from $(g \circ f)(x)$.

An essential aspect of composing functions is the determination of the domain of the composite function. Specifically, the domain of $f \circ g$ consists of all x-values within the domain of g such that the corresponding g(x) lies within the domain of f.

For instance, consider the functions
$$f(x) = 3x + 2 \quad \text{and} \quad g(x) = x^2.$$

The composition $f \circ g$ is determined by replacing x in f(x) with g(x), yielding
$$(f \circ g)(x) = f(x^2) = 3x^2 + 2.$$

In contrast, composing in the reverse order gives
$$(g \circ f)(x) = g(3x + 2) = (3x + 2)^2.$$

The differences between $f \circ g$ and $g \circ f$ illustrate how function composition can generate distinctly different relationships even when the original functions remain the same.

The procedure of composing functions involves careful substitution and verification of appropriate domains. When performing these operations, attention is devoted to ensuring that each function's input requirement is satisfied, thereby yielding valid expressions for the resulting composite function.

Through these operations, functions can be combined in diverse ways, providing a powerful framework for modeling various relationships and enabling the construction of more complex functions from simpler components.

Multiple Choice Questions

1. Which of the following expressions correctly represents the sum of two functions f(x) and g(x)?

    (a) f(x) · g(x)
    (b) f(x) + g(x)
    (c) f(x) - g(x)

(d) f(g(x))

2. If $f(x) = 2x + 3$ and $g(x) = x^2$, what is $(f + g)(x)$?

    (a) $2x + 3x^2$
    (b) $2x + 3 + x^2$
    (c) $(2x + 3)x^2$
    (d) $2x + 3 - x^2$

3. Given $f(x) = x - 1$ and $g(x) = x + 2$, which of the following correctly represents the quotient $\left(\frac{f}{g}\right)(x)$?

    (a) $(x - 1) \cdot (x + 2)$
    (b) $\frac{x+2}{x-1}$
    (c) $\frac{x-1}{x+2}$
    (d) $x - 1 + x + 2$

4. Which of the following definitions best describes the composition of functions $(f \circ g)(x)$?

    (a) $(f \circ g)(x) = f(x) \cdot g(x)$
    (b) $(f \circ g)(x) = f(g(x))$
    (c) $(f \circ g)(x) = f(x) + g(x)$
    (d) $(f \circ g)(x) = g(f(x))$

5. Why is it generally true that $(f \circ g)(x)$ is not equal to $(g \circ f)(x)$?

    (a) Because multiplication of functions is commutative
    (b) Because the order of applying functions affects the outcome
    (c) Because addition of functions is non-associative
    (d) Because division in functions is undefined for all x

6. The domain of the composite function $(f \circ g)(x)$ is best described as:

    (a) All x in the domain of f
    (b) All x in the domain of g
    (c) All x in the domain of g for which $g(x)$ is in the domain of f

(d) The intersection of the domains of f and g

7. For $f(x) = 3x + 2$ and $g(x) = x^2$, what are the expressions for $(f \circ g)(x)$ and $(g \circ f)(x)$, respectively?

(a) $3x^2 + 2$ and $(3x + 2)^2$

(b) $(3x + 2)^2$ and $3x^2 + 2$

(c) $3x + 2x^2$ and $\frac{3x+2}{x^2}$

(d) $3x^2 - 2$ and $(3x - 2)^2$

Answers:

1. **B:** $f(x) + g(x)$
   When adding functions, we define the sum by adding their outputs for each x: $(f + g)(x) = f(x) + g(x)$.

2. **B:** $2x + 3 + x^2$
   Since $(f + g)(x) = f(x) + g(x)$, substituting $f(x) = 2x + 3$ and $g(x) = x^2$ yields $2x + 3 + x^2$.

3. **C:** $\frac{x-1}{x+2}$
   The quotient of two functions is defined by $\left(\frac{f}{g}\right)(x) = \frac{f(x)}{g(x)}$. Thus, with $f(x) = x - 1$ and $g(x) = x + 2$, the expression is $\frac{x-1}{x+2}$.

4. **B:** $(f \circ g)(x) = f(g(x))$
   Composition involves substituting the output of g(x) into f(x), which is exactly represented by $f(g(x))$.

5. **B: Because the order of applying functions affects the outcome**
   In composition, changing the order alters the input to each function. Generally, $f(g(x)) \neq g(f(x))$.

6. **C: All x in the domain of g for which $g(x)$ is in the domain of f**
   The composite function $(f \circ g)(x)$ is defined only for x in the domain of g that yield values $g(x)$ lying within the domain of f.

7. **A:** $3x^2 + 2$ and $(3x + 2)^2$
   For $f(x) = 3x+2$ and $g(x) = x^2$: $(f \circ g)(x) = f(g(x)) = 3x^2 + 2$, and $(g \circ f)(x) = g(f(x)) = (3x + 2)^2$. These expressions illustrate that the order of composition matters.

Practice Problems

1. Let
$$f(x) = 2x + 3 \quad \text{and} \quad g(x) = x^2.$$
Write the expressions for
$$(f + g)(x) \quad \text{and} \quad (f - g)(x),$$
and state the domain of each resulting function.

2. Given the functions
$$f(x) = x - 1 \quad \text{and} \quad g(x) = x + 2,$$
compute the product
$$(f \cdot g)(x) = f(x) \cdot g(x),$$
and the quotient
$$\left(\frac{f}{g}\right)(x) = \frac{f(x)}{g(x)},$$
including any necessary restrictions on the domain.

3. For
$$f(x) = 3x + 2 \quad \text{and} \quad g(x) = x^2,$$
determine the composite function
$$(f \circ g)(x) = f(g(x)),$$
and specify its domain.

4. Let
$$f(x) = \sqrt{x} \quad \text{and} \quad g(x) = x - 4.$$
Find the composite function
$$(g \circ f)(x) = g(f(x)),$$
and determine its domain, taking into account the domain restrictions due to the square root.

5. Consider
$$f(x) = \frac{1}{x} \quad \text{and} \quad g(x) = x - 3.$$
Compute the composite functions
$$(f \circ g)(x) \quad \text{and} \quad (g \circ f)(x),$$

and explain why these two compositions are generally not equal.

6. In a real-world scenario, suppose the temperature in Celsius at x hours after midnight is given by

$$f(x) = 0.5x + 10,$$

and the conversion from Celsius to Fahrenheit is given by

$$g(x) = \frac{9}{5}x + 32.$$

(a) Find the composite function

$$(g \circ f)(x) = g(f(x))$$

that converts the temperature from Celsius (as modeled by $f(x)$) directly into Fahrenheit.

(b) Using your composite function, determine the temperature in Fahrenheit at 6:00 AM.

Answers

1. **Solution:**
   We start with the given functions:
   $$f(x) = 2x + 3 \quad \text{and} \quad g(x) = x^2.$$
   The sum of the two functions is:
   $$(f + g)(x) = f(x) + g(x) = 2x + 3 + x^2.$$
   It is customary to write the result in descending order, so:
   $$(f + g)(x) = x^2 + 2x + 3.$$
   The difference is:
   $$(f - g)(x) = f(x) - g(x) = 2x + 3 - x^2,$$
   or equivalently written as:
   $$(f - g)(x) = -x^2 + 2x + 3.$$
   Since both $f(x)$ and $g(x)$ are defined for all real numbers (they are polynomials), the domain of both $(f + g)(x)$ and $(f - g)(x)$ is all real numbers.

2. **Solution:**
   For the functions:
   $$f(x) = x - 1 \quad \text{and} \quad g(x) = x + 2,$$
   the product is calculated as follows:
   $$(f \cdot g)(x) = f(x) \cdot g(x) = (x - 1)(x + 2).$$
   Expanding this product:
   $$(f \cdot g)(x) = x^2 + 2x - x - 2 = x^2 + x - 2.$$
   The quotient is:
   $$\left(\frac{f}{g}\right)(x) = \frac{x - 1}{x + 2}.$$
   For the quotient, we must ensure the denominator is not zero, thus:
   $$x + 2 \neq 0 \quad \Rightarrow \quad x \neq -2.$$
   Therefore, the domain of $\left(\frac{f}{g}\right)(x)$ is all real numbers except $x = -2$. The product $(f \cdot g)(x)$ is defined for all real numbers.

3. **Solution:**
   Given
   $$f(x) = 3x + 2 \quad \text{and} \quad g(x) = x^2,$$
   the composite function $(f \circ g)(x)$ is computed by substituting $g(x)$ into f:
   $$(f \circ g)(x) = f(g(x)) = f(x^2) = 3x^2 + 2.$$
   Since $g(x) = x^2$ and $f(x) = 3x + 2$ are both polynomials defined for all real numbers, the domain of the composite function is all real numbers.

4. **Solution:**
   With
   $$f(x) = \sqrt{x} \quad \text{and} \quad g(x) = x - 4,$$
   the composite function $(g \circ f)(x)$ is given by:
   $$(g \circ f)(x) = g(f(x)) = g(\sqrt{x}) = \sqrt{x} - 4.$$
   However, because $f(x) = \sqrt{x}$ requires that $x \geq 0$, the domain of f (and hence of $g \circ f$) is $x \geq 0$. The function $g(x) = x - 4$ is defined for all real numbers, so no further restrictions apply. Therefore, the domain of $(g \circ f)(x)$ is $x \geq 0$.

5. **Solution:**
   For the functions:
   $$f(x) = \frac{1}{x} \quad \text{and} \quad g(x) = x - 3,$$
   the composite functions are computed as follows.
   First, compute $(f \circ g)(x)$:
   $$(f \circ g)(x) = f(g(x)) = f(x - 3) = \frac{1}{x - 3}.$$
   For this function, the denominator must not be zero:
   $$x - 3 \neq 0 \quad \Rightarrow \quad x \neq 3.$$
   Next, compute $(g \circ f)(x)$:
   $$(g \circ f)(x) = g(f(x)) = g\left(\frac{1}{x}\right) = \frac{1}{x} - 3.$$

426

Here, the domain comes from $f(x)$, which requires $x \neq 0$, and there is no additional restriction from g.

Comparing the two results:
$$(f \circ g)(x) = \frac{1}{x-3} \quad \text{and} \quad (g \circ f)(x) = \frac{1}{x} - 3.$$

In general, these two expressions are not equal because the operations are performed in different orders. In $(f \circ g)(x)$ the subtraction occurs before taking the reciprocal, while in $(g \circ f)(x)$ the reciprocal is taken before subtracting 3. This non-commutativity is a key property of function composition.

6. **Solution:**
   We are given a temperature function in Celsius:
   $$f(x) = 0.5x + 10,$$
   where x represents the number of hours after midnight. The conversion function from Celsius to Fahrenheit is:
   $$g(x) = \frac{9}{5}x + 32.$$

   (a) Finding the composite function:
   The composite function $(g \circ f)(x)$ is obtained by substituting $f(x)$ into g:
   $$(g \circ f)(x) = g(f(x)) = g(0.5x + 10) = \frac{9}{5}(0.5x + 10) + 32.$$

   Distribute $\frac{9}{5}$:
   $$\frac{9}{5}(0.5x) = \frac{9}{10}x \quad \text{and} \quad \frac{9}{5}(10) = 18.$$

   Thus,
   $$(g \circ f)(x) = \frac{9}{10}x + 18 + 32 = \frac{9}{10}x + 50.$$

   (b) Evaluating at $x = 6$:
   To find the temperature in Fahrenheit at 6:00 AM, substitute $x = 6$:
   $$(g \circ f)(6) = \frac{9}{10}(6) + 50.$$

   Compute:
   $$\frac{9}{10}(6) = \frac{54}{10} = 5.4.$$

Then,
$$(g \circ f)(6) = 5.4 + 50 = 55.4.$$

Therefore, the temperature at 6:00 AM in Fahrenheit is approximately 55.4°F.

Chapter 39

Exponential Functions: Properties and Graphs

Definition and Basic Structure

Exponential functions are defined by the form
$$f(x) = a\,b^x,$$
where a is a nonzero constant representing the initial value and b is a positive constant different from 1 that serves as the base. The variable x may assume any real number, which establishes the domain of the function as all real numbers. The parameter a determines the starting point on the y-axis, identified by $f(0) = a$, while the base b governs the behavior of the function. When $b > 1$, the function exhibits rapid increase, and when $0 < b < 1$, it displays a tendency to diminish. The graph of any exponential function includes a horizontal asymptote at $y = 0$, a line that the function approaches but never reaches.

Growth and Decay Properties

Exponential functions reveal markedly different behaviors based on the value of the base b. For functions in which $b > 1$, the value of b^x increases dramatically as the exponent x increases. This rapid increase, known as exponential growth, is characterized by a function that ascends steeply and unboundedly as x approaches

positive infinity. Formally,
$$\lim_{x \to \infty} a\,b^x = \infty \quad (\text{assuming } a > 0),$$
while for large negative values of x the function approaches zero,
$$\lim_{x \to -\infty} a\,b^x = 0.$$

In contrast, if the base satisfies $0 < b < 1$, the function exhibits exponential decay. In this scenario, as x increases, the factor b^x becomes progressively smaller, causing the overall function to decline toward zero. The rapid decrease in the value of the function contrasts with the unbounded growth observed when $b > 1$. The properties of an exponential function—its domain, the position of its horizontal asymptote, and the nature of its increase or decrease—are determined entirely by the values of a and b.

Graphing Techniques for Exponential Functions

Graphing an exponential function accurately requires the identification of key features that dictate its shape and behavior. The process involves several systematic steps.

1 Identification of Key Characteristics

The initial characteristic of an exponential function is the y-intercept, provided by evaluating the function at $x = 0$:
$$f(0) = a\,b^0 = a.$$
This point serves as an anchor for the graph. The base b defines the rate at which the function increases or decreases; a base greater than 1 results in a graph that rises steeply as x increases, while a base between 0 and 1 causes the graph to fall as x increases. A permanent feature of the graph is the horizontal line $y = 0$, which functions as an asymptote that the graph approaches but never crosses.

2 Constructing the Graph

A detailed graph construction involves evaluating the function at a series of strategically chosen x-values that include negative, zero, and positive values. By calculating and plotting the points corresponding to these x-values, a sequence of points is obtained that illustrates the behavior of the function. The smooth and continuous connection of these points forms the curve of the exponential function. In cases of exponential growth, successive points demonstrate a rapid rise as the exponent increases, whereas in exponential decay the points progressively approach the horizontal axis as x increases.

3 Asymptotic Behavior and Domain Considerations

The domain of any exponential function $f(x) = a\,b^x$ is the set of all real numbers, and this fact is reflected in its graph. The depiction of the horizontal asymptote at $y = 0$ is crucial; it visually reinforces that as x becomes very large in either the positive or negative direction, the function values tend toward, but do not reach, zero. The demonstration of this asymptotic behavior, along with the precise plotting of the y-intercept and the verification of growth or decay tendencies, ensures that the graphical representation accurately mirrors the intrinsic properties of the function.

The comprehensive analysis of exponential functions through their algebraic structure, behavior under growth and decay conditions, and methodical graph construction establishes a robust framework for understanding their properties and applications.

Multiple Choice Questions

1. What is the general form of an exponential function?

    (a) $f(x) = ax + b$
    (b) $f(x) = a\,b^x$, where $a \neq 0$ and b is positive with $b \neq 1$
    (c) $f(x) = a\,x^b$
    (d) $f(x) = a\,\log_b(x)$

2. For the exponential function $f(x) = a\,b^x$, what is the value of $f(0)$?

    (a) 0
    (b) b
    (c) a

(d) $a + b$

3. Which condition on the base b causes the function $f(x) = a\,b^x$ to exhibit exponential growth?

    (a) $0 < b < 1$
    (b) $b < 0$
    (c) $b > 1$
    (d) $b = 1$

4. What is the horizontal asymptote of any exponential function of the form $f(x) = a\,b^x$?

    (a) $x = 0$
    (b) $y = a$
    (c) $y = 0$
    (d) $x = a$

5. For an exponential function with a base satisfying $0 < b < 1$, what best describes the behavior of the function as x increases?

    (a) It grows without bound.
    (b) It decays toward zero.
    (c) It oscillates between positive and negative values.
    (d) It remains constant.

6. Which of the following steps is essential when graphing an exponential function?

    (a) Only plot the y-intercept and horizontal asymptote.
    (b) Plot several points for a mix of negative, zero, and positive x-values, then connect them smoothly.
    (c) Use only the derivative to sketch the curve.
    (d) Focus solely on the behavior as x approaches infinity.

7. In the function $f(x) = a\,b^x$, what role does the coefficient a play?

    (a) It determines the horizontal asymptote.
    (b) It sets the initial value or y-intercept, since $f(0) = a$.

(c) It controls the rate of growth or decay.

(d) It restricts the domain to positive values.

Answers:

1. **B:** The general form of an exponential function is given by $f(x) = a\,b^x$. Here, a is a nonzero constant and b is a positive constant not equal to 1, which distinguishes exponential behavior from linear or logarithmic forms.

2. **C:** Since any nonzero number raised to the zero power is 1, we have $f(0) = a\,(b^0) = a$. This establishes the y-intercept of the graph at the point $(0, a)$.

3. **C:** Exponential growth occurs when $b > 1$. In this case, as x increases, the term b^x increases rapidly, causing the function to grow exponentially.

4. **C:** The horizontal asymptote of an exponential function $f(x) = a\,b^x$ is $y = 0$. Regardless of the values of a and b, the function approaches zero as x tends toward negative infinity (for $b > 1$) or as x tends toward positive infinity (for $0 < b < 1$).

5. **B:** When $0 < b < 1$, the function exhibits exponential decay. As x increases, the factor b^x becomes smaller and the overall function values approach zero.

6. **B:** An effective method for graphing an exponential function is to calculate and plot several points—including points for negative, zero, and positive x values—and then connect them smoothly. This process accurately reflects the function's rapid increase or decay and its asymptotic behavior.

7. **B:** The coefficient a in the function $f(x) = a\,b^x$ gives the initial value, specifically the y-intercept, because $f(0) = a$. It indicates where the graph crosses the y-axis.

Practice Problems

1. For the exponential function
$$f(x) = 3 \cdot 2^x,$$

identify the y-intercept and the horizontal asymptote. Then, describe the end behavior of the function as x approaches positive and negative infinity.

2. Consider the function
$$f(x) = -4 \cdot \left(\frac{1}{2}\right)^x.$$
Determine the y-intercept and horizontal asymptote of this function, and describe its end behavior as x approaches positive and negative infinity.

3. For an exponential function defined by
$$f(x) = a \cdot b^x,$$
if it is given that $f(0) = 6$ and $f(2) = 24$, determine the values of a and b. Provide a detailed explanation of your reasoning.

4. Describe the systematic steps you would take to graph an exponential function such as
$$f(x) = 2^x,$$
and explain how you would identify its key features including the y-intercept and the horizontal asymptote.

5. For the function
$$f(x) = -3 \cdot 4^x,$$
determine the domain and range. Additionally, explain how the negative coefficient affects the graph of the function.

6. Explain why the horizontal asymptote of any exponential function
$$f(x) = a \cdot b^x,$$
with $b > 0$ and $b \neq 1$ is never crossed by its graph. Use limits to justify your explanation.

Answers

1. **Solution:** For the function
$$f(x) = 3 \cdot 2^x,$$
the y-intercept is found by evaluating the function at $x = 0$:
$$f(0) = 3 \cdot 2^0 = 3 \cdot 1 = 3.$$
Therefore, the y-intercept is at $(0, 3)$.

The horizontal asymptote of any exponential function of the form $f(x) = a \cdot b^x$ is $y = 0$ because no matter how large or small x becomes, the exponential portion 2^x never reaches zero—it only approaches it.

For the end behavior:

- As $x \to \infty$: 2^x grows without bound, so
$$\lim_{x \to \infty} 3 \cdot 2^x = \infty.$$

- As $x \to -\infty$: 2^x approaches zero, so
$$\lim_{x \to -\infty} 3 \cdot 2^x = 3 \cdot 0 = 0.$$

Thus, the function increases without bound for large positive x and approaches zero (the horizontal asymptote) for large negative x.

2. **Solution:** For the function
$$f(x) = -4 \cdot \left(\frac{1}{2}\right)^x,$$
we evaluate the y-intercept by setting $x = 0$:
$$f(0) = -4 \cdot \left(\frac{1}{2}\right)^0 = -4 \cdot 1 = -4.$$
So the y-intercept is $(0, -4)$.

The horizontal asymptote is determined by the behavior of the exponential part. Since $\left(\frac{1}{2}\right)^x$ is always positive and never reaches zero, the asymptote is at
$$y = 0.$$

For the end behavior:

- As $x \to \infty$: $\left(\frac{1}{2}\right)^x$ tends to 0. Therefore,
$$\lim_{x \to \infty} -4 \cdot \left(\frac{1}{2}\right)^x = -4 \cdot 0 = 0,$$
but the approach is from the negative side because the constant is negative.
- As $x \to -\infty$: $\left(\frac{1}{2}\right)^x = (2)^{-x}$ grows without bound, so
$$\lim_{x \to -\infty} -4 \cdot \left(\frac{1}{2}\right)^x = -4 \cdot \infty = -\infty.$$

In summary, $f(x)$ approaches zero from below as $x \to \infty$ and decreases without bound as $x \to -\infty$.

3. **Solution:** We are given an exponential function in the form
$$f(x) = a \cdot b^x,$$
with the conditions $f(0) = 6$ and $f(2) = 24$.
First, evaluate $f(0)$:
$$f(0) = a \cdot b^0 = a \cdot 1 = a.$$
Since $f(0) = 6$, we have:
$$a = 6.$$

Next, use $f(2)$:
$$f(2) = 6 \cdot b^2 = 24.$$
Solve for b^2:
$$b^2 = \frac{24}{6} = 4.$$
Because $b > 0$ and $b \neq 1$, the value of b is:
$$b = 2.$$

Therefore, the parameters are $a = 6$ and $b = 2$.

4. **Solution:** To graph an exponential function such as
$$f(x) = 2^x,$$
follow these systematic steps:

(a) **Determine the y-intercept:** Evaluate $f(0)$ to get:
$$f(0) = 2^0 = 1.$$
This point $(0, 1)$ serves as the anchor of the graph.

(b) **Identify the horizontal asymptote:** For any function of the form $f(x) = a \cdot b^x$, the exponential term never equals zero; hence, the horizontal asymptote is at:
$$y = 0.$$

(c) **Plot additional points:** Choose several values of x (including negative, zero, and positive values) to compute $f(x)$. For example:
$$f(-2) = 2^{-2} = \frac{1}{4}, \quad f(-1) = 2^{-1} = \frac{1}{2},$$
$$f(1) = 2^1 = 2, \quad f(2) = 2^2 = 4.$$
These points help illustrate how the function behaves.

(d) **Sketch the curve:** Plot the computed points on the Cartesian plane and draw a smooth curve through them while ensuring it approaches the horizontal asymptote $y = 0$ as $x \to -\infty$ and increases rapidly for $x \to \infty$.

(e) **Analyze their significance:** Recognizing the quick rise of the graph for increasing x and its gradual approach to zero for decreasing x reinforces the concepts of exponential growth and the behavior dictated by the base.

5. **Solution:** For the function
$$f(x) = -3 \cdot 4^x,$$
consider the following aspects:

Domain: Exponential functions are defined for all real numbers regardless of the constants involved. Thus, the domain is:
$$(-\infty, \infty).$$

Range: The basic exponential function 4^x is always positive, meaning $4^x > 0$ for all x. However, the negative coefficient

-3 reflects the graph across the x-axis. Therefore, $f(x)$ is always negative:
$$f(x) < 0.$$
The range is:
$$(-\infty, 0).$$

Effect of the Negative Coefficient: Multiplying by -3 reverses the orientation of the graph relative to the x-axis. While a positive coefficient would result in an exponential graph that lies entirely above the x-axis, the negative coefficient causes the graph to lie below it. The y-intercept is computed as:
$$f(0) = -3 \cdot 4^0 = -3 \cdot 1 = -3,$$
confirming that it starts below the x-axis.

6. **Solution:** For any exponential function of the form
$$f(x) = a \cdot b^x,$$
where $b > 0$ and $b \neq 1$, the horizontal asymptote is $y = 0$. This occurs because the exponential term b^x is always positive and never equals zero for any finite value of x.

To justify this using limits, consider the two cases:

Case 1: $b > 1$ (Exponential Growth) As $x \to -\infty$,
$$\lim_{x \to -\infty} b^x = 0,$$
so
$$\lim_{x \to -\infty} a \cdot b^x = a \cdot 0 = 0.$$
Although when $x \to \infty$ the function grows without bound, the approach to zero on one side confirms the horizontal asymptote.

Case 2: $0 < b < 1$ (Exponential Decay) As $x \to \infty$,
$$\lim_{x \to \infty} b^x = 0,$$
hence
$$\lim_{x \to \infty} a \cdot b^x = a \cdot 0 = 0.$$

In both cases, the output of the exponential part never reaches zero; it only approaches it. Therefore, the horizontal line $y = 0$ is an asymptote that the graph gets arbitrarily close to but never crosses.

Chapter 40

Logarithmic Functions and Their Properties

Definition and Inverse Relationship with Exponential Functions

A logarithmic function arises as the inverse of an exponential function. In its defining relation, for a given base b such that $b > 0$ and $b \neq 1$, the exponential function is characterized by

$$f(x) = b^x.$$

The logarithmic function, denoted by $\log_b x$, is defined in such a manner that the equality

$$y = \log_b x \iff b^y = x$$

holds for all $x > 0$ and $y \in \mathbb{R}$. This mutual inversion is confirmed by the identities

$$b^{\log_b x} = x \quad \text{and} \quad \log_b(b^x) = x,$$

which underscore the precise manner in which the logarithmic function undoes the effect of exponentiation. The domain of $\log_b x$ is the set of all positive real numbers, while its range comprises all real numbers.

Fundamental Laws Governing Logarithms

The algebraic properties of logarithms stem directly from the corresponding properties of exponents. These properties serve as essential tools for simplifying expressions and solving equations that involve logarithms.

1 The Product Rule

For any two positive real numbers M and N, the logarithm of their product can be decomposed into the sum of the logarithms of each factor:
$$\log_b(MN) = \log_b M + \log_b N.$$
This rule is a direct consequence of the exponentiation property $b^{p+q} = b^p \cdot b^q$, and it provides a systematic technique for handling multiplicative structures within logarithmic expressions.

2 The Quotient Rule

When considering the quotient of two positive real numbers M and N, the logarithm adheres to the formula
$$\log_b\left(\frac{M}{N}\right) = \log_b M - \log_b N.$$
The derivation of this rule relies on the exponential identity $\frac{b^p}{b^q} = b^{p-q}$. This property facilitates the simplification of logarithms when division is present.

3 The Power Rule

For any positive real number M and any real exponent p, the logarithm of M raised to the power p is given by
$$\log_b(M^p) = p \cdot \log_b M.$$
This property follows naturally from the exponential rule $(b^x)^p = b^{xp}$ and is instrumental in isolating exponents when solving logarithmic equations.

4 The Change-of-Base Formula

The change-of-base formula allows the conversion of a logarithm from one base to another and is expressed as

$$\log_b M = \frac{\log_k M}{\log_k b},$$

where k is any positive number distinct from 1. This formula proves particularly useful when evaluating logarithms on devices that inherently compute logarithms in a fixed base, such as base 10 or the natural base e.

Graphical Characteristics of Logarithmic Functions

The graph of a logarithmic function of the form

$$f(x) = \log_b x$$

exhibits several distinctive features. The domain of this function is $(0, \infty)$, reflecting the impossibility of taking logarithms of non-positive values. The range is $(-\infty, \infty)$, and the function displays a vertical asymptote along the line $x = 0$. In the case where $b > 1$, the function is monotonically increasing, whereas when $0 < b < 1$ it is monotonically decreasing. The characteristic shape of the graph arises from the fact that logarithmic functions represent the inverse of exponential functions; consequently, their graphs are precise reflections of their exponential counterparts across the line $y = x$.

Algebraic Manipulations Involving Logarithms

The laws of logarithms provide a robust framework for the simplification and manipulation of logarithmic expressions. The properties outlined above allow complex expressions to be rewritten in a more manageable form. For example, an expression such as

$$\log_b \left(\frac{M^p \cdot N^q}{P^r} \right)$$

can be expanded using the product, quotient, and power rules to yield
$$p \log_b M + q \log_b N - r \log_b P.$$
This transformation is particularly advantageous when solving equations or analyzing the behavior of logarithmic expressions in a systematic manner.

Multiple Choice Questions

1. Which of the following is a necessary condition for the base in the logarithmic function $\log_b x$ to be defined?

    (a) b can be any nonzero number.
    (b) b must be greater than 1.
    (c) b must be positive and not equal to 1.
    (d) b must be an integer.

2. The product rule for logarithms expresses the logarithm of a product as:

    (a) $\log_b(MN) = \log_b M \cdot \log_b N$
    (b) $\log_b(MN) = \log_b M + \log_b N$
    (c) $\log_b(MN) = \log_b M - \log_b N$
    (d) $\log_b(MN) = \frac{\log_b M}{\log_b N}$

3. Which of the following correctly represents the power rule for logarithms?

    (a) $\log_b(M^p) = p \cdot \log_b M$
    (b) $\log_b(M^p) = \log_b(p \cdot M)$
    (c) $\log_b(M^p) = (\log_b M)^p$
    (d) $\log_b(M^p) = \log_b M^p$

4. The change-of-base formula allows a logarithm to be expressed in terms of logarithms with another base. Which of the following is the correct change-of-base formula?

    (a) $\log_b M = \frac{\log_k M}{\log_k b}$ where $k > 0$ and $k \neq 1$
    (b) $\log_b M = \frac{\ln M}{\ln b}$

(c) $\log_b M = \frac{\log_b M}{\log_b b}$

(d) $\log_b M = \ln M - \ln b$

5. What is the value of $b^{\log_b x}$ for $x > 0$?

   (a) $b \cdot x$
   (b) x
   (c) $\log_b x$
   (d) $\frac{1}{x}$

6. Which of the following is a correct graphical characteristic of the function $f(x) = \log_b x$ when $b > 1$?

   (a) It is defined for all real numbers x.
   (b) It has a horizontal asymptote at $x = 0$.
   (c) It passes through the point $(1, 0)$ and is increasing.
   (d) It is a decreasing function.

7. Using the laws of logarithms, the expression

$$\log_b \left(\frac{M^2 \cdot N}{P} \right)$$

can be simplified to:

   (a) $2 \log_b M + \log_b N + \log_b P$
   (b) $2 \log_b M + \log_b N - \log_b P$
   (c) $\log_b M^2 + \log_b N + \log_b P$
   (d) $\log_b M^2 - \log_b N - \log_b P$

Answers:

1. **C: b must be positive and not equal to 1**
   For the logarithmic function $\log_b x$ to be defined, the base b must satisfy $b > 0$ and $b \neq 1$. This ensures the function is one-to-one and properly inverses the exponential function.

2. **B: $\log_b(MN) = \log_b M + \log_b N$**
   The product rule for logarithms states that the logarithm of a product equals the sum of the logarithms of the factors. This mirrors the exponent rule $b^{p+q} = b^p \cdot b^q$.

3. **A:** $\log_b(M^p) = p \cdot \log_b M$
   The power rule allows the exponent on the argument of a logarithm to be brought in front as a multiplier. This is a direct consequence of the exponentiation rule $(b^x)^p = b^{xp}$.

4. **A:** $\log_b M = \frac{\log_k M}{\log_k b}$ **where** $k > 0$ **and** $k \neq 1$
   The change-of-base formula is a valuable tool because it converts a logarithm in one base into a quotient of logarithms in any other convenient base, such as 10 or e.

5. **B:** x
   By definition, the logarithmic function is the inverse of the exponential function. Therefore, $b^{\log_b x} = x$ for all $x > 0$.

6. **C: It passes through the point $(1,0)$ and is increasing**
   For $b > 1$, the logarithmic function has a domain of $x > 0$, passes through $(1,0)$ because $\log_b 1 = 0$, has a vertical asymptote at $x = 0$, and is an increasing function.

7. **B:** $2\log_b M + \log_b N - \log_b P$
   Using the logarithm laws: first, $\log_b(M^2) = 2\log_b M$; then, the product rule gives $\log_b(M^2 \cdot N) = 2\log_b M + \log_b N$; finally, the quotient rule subtracts the logarithm of the denominator, yielding $2\log_b M + \log_b N - \log_b P$.

Practice Problems

1. Consider the logarithmic function
$$f(x) = \log_4 x.$$
Determine the value of
$$y = \log_4 64,$$
and then verify the inverse relationship by showing that
$$4^y = 64.$$

2. Simplify the expression
$$\log_2(8 \cdot 16)$$
using the product rule for logarithms, and then evaluate its numerical value.

3. Simplify the expression
$$\log_3\left(\frac{81}{9}\right)$$
using the quotient rule for logarithms, and determine its numerical value.

4. Simplify the expression
$$\log_5(25^3)$$
using the power rule, and then compute its value.

5. Use the change-of-base formula to rewrite the logarithm
$$\log_7 49$$
with base 10, and then evaluate its numerical value.

6. Describe the key features of the graph of
$$f(x) = \log_b x$$
for a base $b > 1$. In your answer, include the domain, range, location of the vertical asymptote, and whether the function is increasing or decreasing.

Answers

1. **Solution:** To find
$$y = \log_4 64,$$
we need to determine the exponent y such that
$$4^y = 64.$$
We note that:
$$4^1 = 4, \quad 4^2 = 16, \quad \text{and} \quad 4^3 = 64.$$

Therefore,
$$y = 3.$$
To verify the inverse relationship, substitute $y = 3$ into 4^y:
$$4^3 = 64,$$
which confirms the original equation. This exercise illustrates the definition of the logarithmic function as the inverse of the exponential function.

2. **Solution:** We are given the expression
$$\log_2(8 \cdot 16).$$
Applying the product rule for logarithms, which states that
$$\log_b(MN) = \log_b M + \log_b N,$$
we write:
$$\log_2(8 \cdot 16) = \log_2 8 + \log_2 16.$$
Now, evaluate each term:
$$\log_2 8 = 3 \quad \text{since } 2^3 = 8,$$
$$\log_2 16 = 4 \quad \text{since } 2^4 = 16.$$
Adding these gives:
$$\log_2(8 \cdot 16) = 3 + 4 = 7.$$
This demonstrates both the proper use of the product rule and the evaluation of the resulting logarithms.

3. **Solution:** We start with the expression
$$\log_3\left(\frac{81}{9}\right).$$
Using the quotient rule for logarithms, which is given by
$$\log_b\left(\frac{M}{N}\right) = \log_b M - \log_b N,$$
we write:
$$\log_3\left(\frac{81}{9}\right) = \log_3 81 - \log_3 9.$$

Next, evaluate each logarithm:
$$\log_3 81 = 4 \quad \text{because } 3^4 = 81,$$
$$\log_3 9 = 2 \quad \text{because } 3^2 = 9.$$
Therefore,
$$\log_3 \left(\frac{81}{9}\right) = 4 - 2 = 2.$$
This solution shows the effective use of the quotient rule and the evaluation of common logarithms.

4. **Solution:** Consider the expression
$$\log_5(25^3).$$
Notice that 25 can be rewritten as 5^2. Thus, we have:
$$25^3 = (5^2)^3 = 5^6.$$
Therefore,
$$\log_5(25^3) = \log_5(5^6).$$
Using the property of logarithms that states
$$\log_b(b^x) = x,$$
we get:
$$\log_5(5^6) = 6.$$
Alternatively, one may use the power rule:
$$\log_5(25^3) = 3 \cdot \log_5 25,$$
and since $\log_5 25 = 2$ (because $25 = 5^2$), it follows that:
$$3 \cdot 2 = 6.$$
Both methods yield the same result, confirming the correct application of the power rule.

5. **Solution:** We are asked to rewrite
$$\log_7 49$$
using the change-of-base formula with base 10. The change-of-base formula is given by:
$$\log_b M = \frac{\log_k M}{\log_k b},$$

where k is the new base (in this case, 10). Hence,
$$\log_7 49 = \frac{\log_{10} 49}{\log_{10} 7}.$$
However, notice that 49 is equal to 7^2. Thus, using the logarithmic identity:
$$\log_7(7^2) = 2,$$
we immediately conclude that:
$$\log_7 49 = 2.$$
The change-of-base formula provides a method for evaluating logarithms on calculators, but recognizing the power relationship here simplifies the evaluation.

6. **Solution:** Consider the graph of the function
$$f(x) = \log_b x,$$
where $b > 1$. The key graphical features include:

Domain: The function is defined only for positive real numbers. Therefore, the domain is
$$(0, \infty).$$

Range: As x varies over all positive numbers, $\log_b x$ can take any real value. Thus, the range is
$$(-\infty, \infty).$$

Vertical Asymptote: The graph has a vertical asymptote at
$$x = 0,$$
because as x approaches 0 from the right, $\log_b x$ decreases without bound (i.e., tends to negative infinity).

Monotonicity: Since the function is the inverse of an exponential function with base $b > 1$, it is monotonically increasing. This means that as x increases, $f(x)$ also increases.

In summary, the graph of
$$f(x) = \log_b x \quad (b > 1)$$
is defined for $x > 0$, spans all real numbers in its range, has a vertical asymptote at $x = 0$, and is continuously increasing. These characteristics reflect its nature as the inverse of an exponential function.

Chapter 41

Solving Exponential and Logarithmic Equations

Solving Exponential Equations

When equations contain exponential expressions, it is often possible to apply techniques based on the properties of exponents. Two broad methods are employed. In some cases, both sides of the equation can be rewritten so that they have the same base. In other cases, the equation does not lend itself to rewriting with a common base and logarithms are employed to extract the variable from an exponent.

1 Exponential Equations with a Common Base

An equation of the form
$$a^{f(x)} = a^{g(x)}$$
allows the use of the one-to-one property of exponential functions. The property guarantees that if the bases are equal and positive (with the base not equal to 1), then the exponents must be equal. That is,
$$a^{f(x)} = a^{g(x)} \implies f(x) = g(x).$$

For example, consider an equation where both sides are expressed with base 2:
$$2^{3x+2} = 2^{x+6}.$$
Since the bases are the same, it is valid to equate the exponents directly:
$$3x + 2 = x + 6.$$
This simple algebraic equation can subsequently be solved by subtracting x from both sides and isolating x:
$$2x + 2 = 6 \implies 2x = 4 \implies x = 2.$$
This method is particularly efficient when the equation can be manipulated into a form where the same base appears on both sides.

2 Exponential Equations Requiring Logarithms

In some cases the exponential equation cannot be rewritten so that both sides share a common base. When confronted with an equation such as
$$c^{f(x)} = d,$$
where $c > 0$, $c \neq 1$, and d is a positive number that is not an obvious power of c, the application of logarithms proves to be an effective strategy. Taking the logarithm of both sides yields
$$\log\left(c^{f(x)}\right) = \log(d).$$
By applying the power rule of logarithms, which states that
$$\log\left(c^{f(x)}\right) = f(x) \cdot \log c,$$
the equation becomes
$$f(x) \cdot \log c = \log d.$$
This equation can then be solved for x by isolating $f(x)$ and performing the necessary algebraic manipulations. For instance, if the equation is
$$3^{2x-1} = 7,$$
taking logarithms on both sides gives
$$(2x - 1) \cdot \log 3 = \log 7.$$
Solving for x proceeds by first dividing both sides by $\log 3$ and then isolating x:
$$2x - 1 = \frac{\log 7}{\log 3} \implies 2x = \frac{\log 7}{\log 3} + 1 \implies x = \frac{1}{2}\left(\frac{\log 7}{\log 3} + 1\right).$$

Solving Logarithmic Equations

Equations that involve logarithmic expressions require a systematic approach that typically begins with the application of logarithmic properties. The goal is to simplify the given expression and ultimately eliminate the logarithm through exponentiation. A careful analysis of the domain restrictions imposed by the logarithm is essential during this process.

1 Combining and Isolating Logarithmic Expressions

Many logarithmic equations appear with more than one logarithmic term. Through the product, quotient, and power rules of logarithms, it is possible to combine these terms into a single logarithmic expression. For instance, consider the following two rules:

$$\log_b M + \log_b N = \log_b(MN)$$

$$\log_b M - \log_b N = \log_b\left(\frac{M}{N}\right).$$

These properties allow a logarithmic equation such as

$$\log_b(M) + \log_b(N) = C,$$

to be rewritten as
$$\log_b(MN) = C.$$

Combining logarithmic expressions in this manner is a crucial step that simplifies the equation and prepares it for conversion from logarithmic to exponential form. It is important to ensure that the arguments of all logarithms remain positive during these operations.

2 Removing the Logarithm through Exponentiation

Once a logarithmic equation has been consolidated to a single logarithm, the logarithm can be eliminated by rewriting the equation in exponential form. Given an equation of the form

$$\log_b(A) = C,$$

the equivalence between logarithmic and exponential expressions leads to
$$A = b^C.$$

This transformation replaces the logarithmic equation with an algebraic equation wherein the variable may now be isolated through standard techniques. When applying this method, the domain restriction that $A > 0$ must be observed. This requirement ensures that the solution obtained is valid in the context of the original logarithmic expression.

Equations Involving Both Exponential and Logarithmic Terms

Certain equations involve a combination of exponential and logarithmic expressions. These mixed equations benefit from the inverse relationship that exists between exponentials and logarithms. The strategy is to manipulate the equation such that either the exponential or the logarithmic component is isolated.

For example, an equation might involve a logarithmic term on one side and an exponential term on the other:

$$a^{f(x)} = \log_b(g(x)) + h.$$

One approach is to isolate the logarithmic term before applying exponentiation to eliminate it. Alternatively, if the logarithmic term remains embedded within a more complex expression, it may be more advantageous to isolate the exponential term and then apply logarithms to both sides. In either case, recognition of the mathematically inverse nature of the functions is essential.

These procedures typically rely on the following properties:

$$b^{\log_b(x)} = x,$$

and

$$\log_b(b^x) = x.$$

By using these identities, an equation can be transformed step by step until a form is attained that allows the variable to be solved by straightforward algebraic manipulation. Each transformation must adhere strictly to the rules of algebra and the domain restrictions inherent to logarithmic functions.

Throughout the process of solving mixed equations, attention is given to ensuring that each step preserves the integrity of the original equation. This includes verifying that all transformations are reversible and that no extraneous solutions are introduced. Rigorous application of logarithmic and exponential properties yields a clear pathway to isolate and solve for the variable in question.

Multiple Choice Questions

1. Which condition must be met for the one-to-one property of exponential functions to guarantee that if
$$a^{f(x)} = a^{g(x)},$$
then
$$f(x) = g(x)?$$

    (a) The base a can be any nonzero number.
    (b) The base a must be positive and not equal to 1.
    (c) The base a must be an integer.
    (d) The exponents $f(x)$ and $g(x)$ must both be linear.

2. Solve the equation
$$2^{3x+2} = 2^{x+6}$$
by equating the exponents. What is the value of x?

    (a) $x = 1$
    (b) $x = 2$
    (c) $x = 3$
    (d) $x = 4$

3. When faced with an exponential equation of the form
$$c^{f(x)} = d,$$
where d is not an obvious power of c, which method is recommended to solve for x?

    (a) Apply the one-to-one property directly by equating $f(x)$ to d.
    (b) Take the logarithm of both sides and then use the power rule to bring down the exponent.

(c) Recast the equation as a quadratic in x and solve using the quadratic formula.

(d) Change the base of the exponential expression arbitrarily to simplify the equation.

4. Which logarithmic property allows you to combine the expression
$$\log_b M + \log_b N$$
into a single logarithm?

(a) The Power Rule
(b) The Change-of-Base Formula
(c) The Product Rule
(d) The Quotient Rule

5. For the logarithmic equation
$$\log_b(A) = C$$
to be defined, which of the following conditions must be true about A?

(a) A can be any real number.
(b) $A > 0$.
(c) A must be negative.
(d) $A \geq 0$.

6. After combining multiple logarithmic terms into a single expression, you obtain
$$\log_b(MN) = C.$$
What is the next appropriate step to solve for the variable?

(a) Take the logarithm of both sides again.
(b) Rewrite the equation in exponential form as $MN = b^C$.
(c) Subtract MN from both sides.
(d) Factor the expression inside the logarithm.

7. In solving an equation that combines both an exponential term and a logarithmic term, such as
$$a^{f(x)} = \log_b\bigl(g(x)\bigr) + h,$$
what is the best strategy to isolate and solve for x?

(a) Substitute $u = f(x)$ to form a quadratic equation.

(b) Isolate either the exponential term or the logarithmic term first, then apply the appropriate inverse operation.

(c) Equate the exponent directly with the logarithm without any additional manipulation.

(d) Multiply both sides by a constant to eliminate the logarithmic term.

Answers:

1. **B: The base a must be positive and not equal to 1**
   Explanation: An exponential function is one-to-one only if its base is positive and different from 1. These conditions ensure that equal outputs imply equal exponents.

2. **B: $x = 2$**
   Explanation: Since the bases are the same, set the exponents equal: $3x + 2 = x + 6$. Solving this equation gives $2x = 4$ and hence $x = 2$.

3. **B: Take the logarithm of both sides and then use the power rule to bring down the exponent**
   Explanation: When you cannot rewrite both sides with the same base, applying logarithms (using any convenient base) permits the use of the power rule—which turns the exponent into a multiplicative factor—thus enabling you to solve for x.

4. **C: The Product Rule**
   Explanation: The product rule of logarithms states that $\log_b M + \log_b N = \log_b(MN)$, allowing you to combine two logarithmic expressions into one.

5. **B: $A > 0$**
   Explanation: A logarithm is defined only when its argument is strictly positive. Therefore, for $\log_b(A)$ to be defined, A must be greater than 0.

6. **B: Rewrite the equation in exponential form as $MN = b^C$**
   Explanation: Converting the logarithmic equation $\log_b(MN) = C$ into the exponential form $MN = b^C$ is the standard step for eliminating the logarithm and simplifying the equation so that x can be isolated.

7. **B: Isolate either the exponential term or the logarithmic term first, then apply the appropriate inverse operation**
   Explanation: In mixed equations, isolating the exponential or logarithmic part allows you to apply the inverse operation (logarithm or exponentiation, respectively) to that isolated term. This step-by-step approach simplifies the equation and facilitates solving for x.

Practice Problems

1. Solve the exponential equation:
$$3^{2x-1} = 3^{x+2}.$$

2. Solve the exponential equation:
$$2^{3x} = 20.$$

3. Solve the logarithmic equation:
$$\log_2(x) + \log_2(x-2) = 3.$$

4. Solve the logarithmic equation:
$$\log_5(x+1) - \log_5(x-2) = 1.$$

5. Solve the equation:
$$\log_5(2^x + 3) = \log_5(7).$$

6. Solve the equation:
$$\ln\left(e^{2x} + 1\right) = \ln\left(3e^x\right).$$

Answers

1. **Solve the exponential equation:**

$$3^{2x-1} = 3^{x+2}.$$

   Solution: Since both sides of the equation have the same base (3), and because 3 is positive and not equal to 1, the one-to-one property of exponential functions allows us to equate the exponents directly. Thus,

$$2x - 1 = x + 2.$$

   Subtracting x from both sides gives

$$x - 1 = 2.$$

   Adding 1 to both sides yields

$$x = 3.$$

   Therefore, the solution is $x = 3$.

2. **Solve the exponential equation:**

$$2^{3x} = 20.$$

   Solution: In this case the equation cannot be rewritten so that both sides have the same base, so we apply logarithms to both sides. Taking the logarithm (of any base, but here we use the common logarithm) gives

$$\log\left(2^{3x}\right) = \log(20).$$

Using the power rule for logarithms (which states that $\log(2^{3x}) = 3x \cdot \log(2)$), we have

$$3x \cdot \log(2) = \log(20).$$

Solving for x by dividing both sides by $3\log(2)$ results in

$$x = \frac{\log(20)}{3\log(2)}.$$

Hence, the solution is $x = \frac{\log(20)}{3\log(2)}$.

3. **Solve the logarithmic equation:**

$$\log_2(x) + \log_2(x-2) = 3.$$

Solution: Apply the product rule for logarithms which states that $\log_2(M) + \log_2(N) = \log_2(MN)$. Hence, the equation becomes

$$\log_2[x(x-2)] = 3.$$

Rewriting the logarithmic equation in its exponential form yields

$$x(x-2) = 2^3.$$

Since $2^3 = 8$, we have

$$x(x-2) = 8.$$

Expanding the left-hand side gives

$$x^2 - 2x - 8 = 0.$$

This quadratic equation can be solved using the quadratic formula:

$$x = \frac{2 \pm \sqrt{(-2)^2 - 4(1)(-8)}}{2} = \frac{2 \pm \sqrt{4+32}}{2} = \frac{2 \pm \sqrt{36}}{2} = \frac{2 \pm 6}{2}.$$

Thus, the potential solutions are

$$x = \frac{2+6}{2} = 4 \quad \text{or} \quad x = \frac{2-6}{2} = -2.$$

However, the domain of the logarithm requires that the arguments be positive. Since $\log_2(x)$ is defined only for $x > 0$ and $\log_2(x-2)$ is defined only when $x - 2 > 0$ (that is, $x > 2$), the only acceptable solution is

$$x = 4.$$

4. **Solve the logarithmic equation:**

$$\log_5(x+1) - \log_5(x-2) = 1.$$

Solution: Use the quotient rule for logarithms which states that $\log_b(M) - \log_b(N) = \log_b\left(\frac{M}{N}\right)$ to combine the terms:

$$\log_5\left(\frac{x+1}{x-2}\right) = 1.$$

Converting the logarithmic equation to its exponential form gives

$$\frac{x+1}{x-2} = 5^1 = 5.$$

Multiplying both sides by $x-2$ we obtain

$$x + 1 = 5(x-2).$$

Expanding the right-hand side:

$$x + 1 = 5x - 10.$$

Subtracting x from both sides leads to

$$1 = 4x - 10.$$

Adding 10 to both sides, we have

$$11 = 4x.$$

Finally, dividing by 4 gives

$$x = \frac{11}{4}.$$

A quick check of the domain shows that $x+1 > 0$ and $x-2 > 0$ are satisfied when $x > 2$. Since $\frac{11}{4} \approx 2.75$ is greater than 2, the solution is valid.

5. **Solve the equation:**

$$\log_5(2^x + 3) = \log_5(7).$$

Solution: Since the logarithms have the same base and the logarithmic function is one-to-one, we equate the arguments:

$$2^x + 3 = 7.$$

Subtract 3 from both sides to isolate the exponential part:
$$2^x = 4.$$
Recognize that 4 can be written as 2^2:
$$2^x = 2^2.$$
Therefore, by the one-to-one property of exponential functions,
$$x = 2.$$

6. **Solve the equation:**
$$\ln\left(e^{2x} + 1\right) = \ln(3e^x).$$

Solution: Since the natural logarithm is one-to-one, we can set the arguments equal to each other:
$$e^{2x} + 1 = 3e^x.$$
To simplify the equation, let $y = e^x$, noting that $y > 0$. The equation becomes
$$y^2 + 1 = 3y.$$
Rearranging yields the quadratic equation
$$y^2 - 3y + 1 = 0.$$
Applying the quadratic formula,
$$y = \frac{3 \pm \sqrt{9-4}}{2} = \frac{3 \pm \sqrt{5}}{2}.$$
Since $y = e^x$ must be positive, both solutions are acceptable:
$$y = \frac{3 + \sqrt{5}}{2} \quad \text{and} \quad y = \frac{3 - \sqrt{5}}{2}.$$
Finally, take the natural logarithm of both sides to solve for x:
$$x = \ln\left(\frac{3 + \sqrt{5}}{2}\right) \quad \text{or} \quad x = \ln\left(\frac{3 - \sqrt{5}}{2}\right).$$
Therefore, the solutions are
$$x = \ln\left(\frac{3 + \sqrt{5}}{2}\right) \quad \text{and} \quad x = \ln\left(\frac{3 - \sqrt{5}}{2}\right).$$

Chapter 42

Sequences and Algebraic Patterns

Arithmetic Sequences

1 Definition and Basic Properties

An arithmetic sequence is defined as an ordered set of numbers in which the difference between any two consecutive terms remains constant. This fixed interval, known as the common difference and denoted by d, establishes a linear progression within the sequence. The first term, represented as a_1, together with the common difference, determines every subsequent term. The general or explicit formula for the nth term can be written as

$$a_n = a_1 + (n-1)d,$$

where n is a positive integer. This formulation emphasizes the uniform step-by-step increase or decrease characteristic of arithmetic sequences.

2 Determining the nth Term

The explicit formula for the nth term permits direct computation without sequentially deriving all previous terms. By substituting the desired position n into the formula, the term can be obtained efficiently. This approach allows for a comprehensive analysis of the sequence's behavior and lays the groundwork for solving problems involving linear relationships in algebra.

3 Recursive Definitions in Arithmetic Sequences

In addition to an explicit formula, arithmetic sequences can be defined recursively. Beginning with an initial term a_1, the sequence evolves according to the relation

$$a_{n+1} = a_n + d.$$

This recursive definition highlights the iterative process by which each term in the sequence is generated from its immediate predecessor. Such a presentation underscores the stepwise nature of arithmetic progressions and provides an alternative perspective that is useful in various algebraic contexts.

Geometric Sequences

1 Definition and Fundamental Characteristics

A geometric sequence is characterized by a constant ratio between successive terms. This ratio, referred to as the common ratio and designated by r, defines the multiplicative progression of the sequence. The general formula for the nth term of a geometric sequence is given by

$$a_n = a_1 \cdot r^{(n-1)},$$

where a_1 denotes the first term and n is a positive integer. The exponential appearance of the index in the formula reflects the nature of the sequence, where each term is produced by multiplying the previous term by the common ratio r.

2 Analyzing the nth Term

The explicit formulation for the nth term allows for the rapid evaluation of any term within the sequence. Variations in the value of r influence the behavior of the sequence significantly. When r is greater than one, the sequence exhibits exponential growth, whereas a value between zero and one leads to exponential decay. Should r be negative, the sequence demonstrates an alternating pattern, with the sign of each term changing in accordance with the multiplication by r.

3 Recursive Formulations in Geometric Sequences

Geometric sequences also admit a recursive definition that mirrors their multiplicative structure. With the initial term a_1 established, the subsequent term is defined by the relation

$$a_{n+1} = a_n \cdot r.$$

This recursive approach encapsulates the fundamental operation governing the sequence and offers an instructive method for constructing the sequence term by term, reinforcing the conceptual understanding of geometric progression.

Algebraic Patterns and Recursive Definitions

1 Recognition and Analysis of Algebraic Patterns

Algebraic patterns emerge from consistent operations or relationships applied to elements of a sequence. These patterns may be observed in both arithmetic and geometric sequences, as well as in more complex configurations. The process of identifying an algebraic pattern involves discerning the underlying rule that dictates the formation of each term. Such patterns are often revealed through systematic observation or by employing algebraic techniques that expose recurring structures within the sequence.

2 Establishing Recursive Definitions

Recursive definitions provide an elegant framework for describing sequences wherein each term is defined in terms of one or more preceding terms. This method of definition is not confined to arithmetic or geometric progressions but extends to any sequence that adheres to a consistent operational rule. A recursively defined sequence typically specifies one or more base cases, which serve as the starting point, and a recursive rule that describes how subsequent terms are generated. This approach emphasizes the functional dependence of later terms on earlier ones and aids in understanding the dynamic properties inherent in algebraic sequences.

3 Examples of Recursive Patterns in Algebra

Recursive sequences may involve a combination of operations that produce non-linear progressions and intricate patterns. For instance, a sequence might be defined by a recursive relationship that incorporates both addition and multiplication, resulting in a progression that cannot be easily expressed in a simple explicit formula. The exploration of such recursive patterns enhances the comprehension of algebraic structures and provides insight into the methods used to construct sequences from specified initial conditions and recursive rules. The interplay between explicit formulas and recursive definitions enriches the study of algebraic patterns, establishing a versatile toolkit for analyzing and modeling a wide array of phenomena in algebra.

Multiple Choice Questions

1. What is the explicit formula for the nth term of an arithmetic sequence?

    (a) $a_n = a_1 + nd$
    (b) $a_n = a_1 + (n-1)d$
    (c) $a_n = a_1 \cdot n + d$
    (d) $a_n = a_1 \cdot (n-1) + d$

2. Which recursive formula correctly defines an arithmetic sequence?

    (a) $a_{n+1} = a_n \times d$
    (b) $a_{n+1} = a_n + d$
    (c) $a_{n+1} = a_n - d$
    (d) $a_{n+1} = a_n / d$

3. What is the explicit formula for the nth term of a geometric sequence?

    (a) $a_n = a_1 + (n-1)r$
    (b) $a_n = a_1 \cdot r^n$
    (c) $a_n = a_1 \cdot r^{n-1}$
    (d) $a_n = a_1 \cdot r^{n+1}$

4. Which recursive formula correctly defines a geometric sequence?

   (a) $a_{n+1} = a_n + r$
   (b) $a_{n+1} = a_n \cdot r$
   (c) $a_{n+1} = a_n - r$
   (d) $a_{n+1} = a_n/r$

5. In an arithmetic sequence, which component remains constant throughout?

   (a) The common ratio
   (b) The common difference
   (c) The product of consecutive terms
   (d) The rate of exponential growth

6. If an arithmetic sequence starts with $a_1 = 4$ and has a common difference $d = 3$, what is the 5th term?

   (a) 19
   (b) 16
   (c) 22
   (d) 18

7. Which of the following best describes the purpose of using recursive definitions in sequences?

   (a) They allow you to calculate any term directly without reference to previous terms.
   (b) They provide a step-by-step process where each term is defined based on the term that precedes it.
   (c) They help to simplify complex exponential equations.
   (d) They apply only to arithmetic sequences.

Answers:

1. **B:** $a_n = a_1 + (n-1)d$
   This is the correct formula for the nth term of an arithmetic sequence, reflecting that each term is obtained by adding the common difference d a total of $n-1$ times to the first term a_1.

2. **B:** $a_{n+1} = a_n + d$
   In an arithmetic sequence, each term is produced by adding the common difference d to the previous term. This recursive definition captures that iterative process.

3. **C:** $a_n = a_1 \cdot r^{n-1}$
   For a geometric sequence, every term is found by multiplying the previous term by the constant common ratio r. The explicit formula uses r^{n-1} to represent the cumulative multiplication from the first term a_1.

4. **B:** $a_{n+1} = a_n \cdot r$
   A geometric sequence is characterized by each term being the product of the previous term and the common ratio r. This recursive formula clearly embodies that relationship.

5. **B: The common difference**
   In an arithmetic sequence, the constant value that is added to each term to produce the next one is called the common difference.

6. **B: 16**
   Using the formula for an arithmetic sequence, $a_5 = a_1 + (5-1)d = 4 + 4 \times 3 = 4 + 12 = 16$.

7. **B: They provide a step-by-step process where each term is defined based on the term that precedes it.**
   Recursive definitions emphasize the iterative nature of sequences by establishing one or more base cases and a rule for determining subsequent terms, offering insight into the underlying progression.

Practice Problems

1. Consider an arithmetic sequence with the first term
$$a_1 = 3$$
and a common difference
$$d = 5.$$
Using the explicit formula, find the 8th term of this sequence.

2. Provide the recursive definition for an arithmetic sequence that starts with
$$a_1 = 7$$
and has a common difference
$$d = -2.$$

3. For a geometric sequence with the first term
$$a_1 = 4$$
and a common ratio
$$r = \frac{1}{2},$$
use the explicit formula to find the 6th term.

4. Write the recursive definition for a geometric sequence where
$$a_1 = 10$$
and the common ratio is
$$r = 3.$$

5. Consider a sequence defined recursively by
$$T_1 = 2 \quad \text{and} \quad T_{n+1} = 3T_n + 1 \quad \text{for } n \geq 1.$$
Determine the value of T_3.

6. Given the explicit formula for a sequence
$$a_n = 2n^2 - n + 3,$$
determine whether this sequence is arithmetic, geometric, or neither. Explain your reasoning.

Answers

1. **Solution:**
    The explicit formula for the n-th term of an arithmetic sequence is given by
    $$a_n = a_1 + (n-1)d.$$
    Here, the values are
    $$a_1 = 3, \quad d = 5, \quad n = 8.$$

Substitute these into the formula:
$$a_8 = 3 + (8-1) \cdot 5 = 3 + 7 \cdot 5 = 3 + 35 = 38.$$
Therefore, the 8th term of the sequence is
$$38$$
.

2. **Solution:**
A recursive definition for an arithmetic sequence specifies the first term and describes how to obtain each subsequent term. With the given initial term
$$a_1 = 7,$$
and common difference
$$d = -2,$$
the recursive rule is:
$$a_{n+1} = a_n - 2.$$
This indicates that each term is 2 less than the previous term.

3. **Solution:**
The explicit formula for the n-th term of a geometric sequence is
$$a_n = a_1 \cdot r^{(n-1)}.$$
For the given sequence, we have
$$a_1 = 4, \quad r = \frac{1}{2}, \quad n = 6.$$
Substitute these into the formula:
$$a_6 = 4 \cdot \left(\frac{1}{2}\right)^{6-1} = 4 \cdot \left(\frac{1}{2}\right)^5.$$
Since
$$\left(\frac{1}{2}\right)^5 = \frac{1}{32},$$
it follows that:
$$a_6 = 4 \cdot \frac{1}{32} = \frac{4}{32} = \frac{1}{8}.$$

Thus, the 6th term is
$$\frac{1}{8}$$
.

4. **Solution:**
In a geometric sequence, the recursive definition starts with the initial term and follows a rule where each term is obtained by multiplying the previous term by the common ratio. Here, the initial term is
$$a_1 = 10,$$
and the common ratio is
$$r = 3.$$
Therefore, the recursive formula is:
$$a_{n+1} = 3 \cdot a_n,$$
with
$$a_1 = 10.$$

5. **Solution:**
The sequence is defined recursively by:
$$T_1 = 2, \quad T_{n+1} = 3T_n + 1.$$
To find
$$T_3,$$
we first calculate
$$T_2$$
:
$$T_2 = 3T_1 + 1 = 3 \cdot 2 + 1 = 6 + 1 = 7.$$
Next, use
$$T_2$$
to determine
$$T_3$$
:
$$T_3 = 3T_2 + 1 = 3 \cdot 7 + 1 = 21 + 1 = 22.$$
Hence,
$$T_3 = 22$$
.

6. **Solution:**
The explicit formula for the sequence is:
$$a_n = 2n^2 - n + 3.$$

For a sequence to be arithmetic, the difference between consecutive terms (i.e.,
$$a_{n+1} - a_n$$
) must be constant; for a geometric sequence, the ratio
$$a_{n+1}/a_n$$
must be constant.

Compute the difference:
$$a_{n+1} = 2(n+1)^2 - (n+1) + 3.$$

Expanding
$$a_{n+1}$$
:
$$a_{n+1} = 2(n^2 + 2n + 1) - n - 1 + 3 = 2n^2 + 4n + 2 - n - 1 + 3$$
$$= 2n^2 + 3n + 4.$$

Now, find the difference:
$$a_{n+1} - a_n = (2n^2 + 3n + 4) - (2n^2 - n + 3)$$
$$= 2n^2 + 3n + 4 - 2n^2 + n - 3 = 4n + 1.$$

Since
$$4n + 1$$
depends on
$$n$$
and is not constant, the sequence is not arithmetic.

Next, examine the possibility of it being geometric by considering the ratio:
$$\frac{a_{n+1}}{a_n}.$$

Because the formula is quadratic in
$$n$$

, this ratio will vary with

$$n$$

and will not remain constant.

Therefore, the sequence defined by

$$a_n = 2n^2 - n + 3$$

is neither arithmetic nor geometric. It is an example of a sequence with an algebraic (quadratic) pattern.